T0341098

THE
ENTREPRENEURIAL
PROJECT
MANAGER

Best Practices and Advances in Program Management Series

Series Editor
Ginger Levin

THE
ENTREPRENEURIAL
PROJECT
MANAGER

CHRISTOPHER COOK

CRC Press
Taylor & Francis Group
Boca Raton London New York

CRC Press is an imprint of the
Taylor & Francis Group, an **informa** business
AN AUERBACH BOOK

CRC Press
Taylor & Francis Group
6000 Broken Sound Parkway NW, Suite 300
Boca Raton, FL 33487-2742

© 2017 by Taylor & Francis Group, LLC
CRC Press is an imprint of Taylor & Francis Group, an Informa business

No claim to original U.S. Government works

Printed on acid-free paper

International Standard Book Number-13: 978-1-4987-8235-7 (Hardback)

Visit the Taylor & Francis Web site at
http://www.taylorandfrancis.com

and the CRC Press Web site at
http://www.crcpress.com

Contents

Preface

Starting off as a laborer in the construction industry, I had the opportunity to see projects being run from a boots-on-the-ground perspective. Project managers drive around in their company vehicles, stop to look at a few things, then drive off—not to be seen the rest of the day. I thought to myself, "What an easy job!" Drive around, talk on the phone, look at plans, and do it all again the next day. Project managers were not the ones installing the storm sewer pipe or having to dig holes to expose underground utilities. They could escape the elements and go back to their air-conditioned vehicles or offices. Some of them even had the chance to go somewhere for lunch and enjoy their meal. It seemed like some far-off land.

I decided to make this my goal: How can I get out of these dirty jeans and ragged t-shirts and into a polo shirt and clean khakis? Going to college became a priority. Learning about management while still experiencing the troubles in the field became my advantage. At the same time, I was also learning about how to read plans and build estimates while putting those calculations to practice with my own two hands.

As time went on, I took on more of a leadership role, and more responsibility came my way. I started to get more involved with those project managers driving around in trucks discussing how the task was progressing and giving them feedback on the design. I saw this as an inducement to enhance my education by going to graduate school in project management. The discussions I was having in the field drove me to pursue a greater understanding of the science behind the thought process.

Attending graduate school led to the writing of this book. I was presented the opportunity to write about the project management profession. It was an honor. I started to look around at project managers and see what needed work. I would sit in the meetings, hear them talk, then observe how those messages were translated. I would talk with field personnel to get their feelings on the direction of projects. I began realizing that the message being conveyed and the actions being taken were vastly different. The disconnect was evident, yet actions to bring the sides closer were not present.

These observations and my interest in other fields are the reasons for this book. My interest in sports and the team dynamics of project management have a closer correlation than appears on the surface. Also, my interest in philosophy started to rise. People have dealt with problems throughout history. Problems more severe than a project behind schedule or over budget. Problems that put people's lives at stake. Melding these dynamics together is the goal of this book. Using words from great philosophers such as Marcus Aurelius and Lao Tzu paired with teachings from great coaches such as Bob Knight and Bill Walsh shows how a project manager can apply the same techniques these people did to their situations.

No matter your industry or experience level, this book will give you mind-sets to improve any project. The biggest improvements a project manager can make are about him- or herself. The individual can be the hardest obstacle to overcome. In writing this book, these techniques have been emphasized in my daily life. I find myself realizing how little control I actually have over events. All the right things can be said and done, yet the results are unpredictable. The tools and techniques used throughout the book will give you the ability to handle anything that may come your way.

An entrepreneur looks for opportunities. An entrepreneurial project man-ager becomes problem-seeking. Solving problems proactively puts you on another level from your contemporaries. This book was a learning and growing experience for me. I hope to pass along that same journey to you.

Acknowledgments

I want to thank Dr. Ginger Levin for planting the seed of an idea to write a book I never thought possible; John Wyzalek of Taylor & Francis for continuing to answer my questions and help me with this process; and Theron Shreve, Lynne Lackenbach, and Marje Pollack of DerryField Publishing Services for editing and keeping me on task. This book would not be possible without your help, and I appreciate the hours you have all given me.

About the Author

Over the past ten years, Christopher Cook, PMP, has spent his career in the construction industry. Aspects of the industry include heavy highway, aggregates, sewer, and grading roads. During this time, he earned his Bachelor's of Science in Industrial Technology Management with an emphasis in Building Construction Management, Master's of Science in Project Management, and Project Management Professional (PMP)® certification. In those years, he served as a laborer and worked his way up to estimator and master scheduler roles. Serving in these various roles has provided many opportunities to monitor and control projects from headquarters. The portfolio of projects is in the tens of millions of dollars annually. Working side by side with operations, he has the technical skills necessary for daily work while learning the management way of doing things to serve as a better manager. He is a member of the local Project Management Institute (PMI) chapter in Denver, Colorado. Having the right education and technical background has proven vital to his success and continued growth in the industry. He shares his project management insights weekly through "The EntrePMeur" blog (http://entrepmeur.wpengine.com).

Chapter 1

Introduction

Project managers have the technical skills to handle any project. You have studied the materials, worked in the industries, and provided the guidance for success. My approach is to assume those technical skills and build on your ability to lead a team through philosophies (mindset changes), sales, childlike thinking, adapting, and narrowing down the topics you focus on to take advantage of the time you have. This new approach brings an entrepreneurial spirit to the project management world. You want to take the tools and techniques you have learned over the years and expand them to new heights. Entrepreneurs are constantly changing and adapting to the world around them. They have to stay cutting edge to make their businesses thrive. Staying cutting edge is the approach I want you to take into project management. Instead of business, it will be a project, but the principles will be relatable to any discipline. The goal is to take your technical skills as a project manager, add the elements of an entrepreneur, and create a high-powered team around you as well as becoming the best project manager you can be.

Entrepreneur. A word that describes creative, driven, goal-oriented individuals, usually in a business sense. My goal throughout this book is to bring this word *entrepreneur* to the project management landscape. Joining an organization should not limit the ability for an individual to thrive creatively and to use unique abilities to lead a team to success. The landscape of the corporate world lies in red tape and an inability to move forward quickly. Having an entrepreneurial mindset in that landscape proves difficult because of its limitations and slow processes. The larger the organization, the longer it takes for an initiative to take hold. This situation is where the entrepreneur can come in, find ways around those obstacles, and thrive in an organization that has proven difficult.

Being efficient is not good enough; effectiveness coupled with that efficiency proves greater. Bill Gates once said, "The first rule of any technology used in a business is that automation applied to an efficient operation will magnify the efficiency. The second is that automation applied to an inefficient operation will magnify the inefficiency" (Gates, 2016). The same goes for applying an entrepreneurial mindset. If you get to be extremely efficient at answering emails and getting to inbox zero, but none of those emails pertains to important tasks at hand, you are magnifying the inefficiency. No one wants to add to the problem. Putting your head down and getting to work may be the worst thing you can do for an organization that is realizing its inefficiencies. Bring the creative, solution-oriented stylings of an entrepreneur to that setting and dominate.

Aggressive may be another term used for an entrepreneur, as someone who sees a problem and is not afraid to tackle it immediately. Later on, I will discuss ways to identify a problem, find its solution, and become the go-to person for solving problems. How valuable would that be for your organization? Having a group of individuals actively seeking problems to solve instead of running from them and looking for others to step up and solve them for you. That is exactly what an entrepreneur brings to the table. Entrepreneurs can use the vast resources of an organization along with their aggressive attitudes toward problems to create a completely different atmosphere. Getting these people into the project management role and leading teams can have deep impacts throughout an organization.

Part of the entrepreneurial spirit dives into the depths of the person. Philosophy is a key aspect of knowing not only your team but who you are. Going 100 miles an hour and not stopping to smell the roses has impacts on relationships within your team and organization. While hard work is the desired attribute of any project manager, there has to be a time for reflection. Is this the right direction for the organization? Is the team working on projects that satisfy the goals of the organization? Am I doing the best I can at leading this team? Do I actively seek out problems to solve, or am I being reactive? These questions and many more are involved with the deeper essence of what a project manager needs to do. Being the first to show up and the last to leave are great characteristics, but if they are misdirected, what good does that do? Allowing yourself those moments of reflection can prove to be a multiplier for results.

Some ways to reflect are through meditation or the reading of philosophy. The philosophies I choose to illustrate project management techniques are Stoicism and Taoism. They lead to practical applications and are not built on whimsical tales. Philosophy has a connotation for being boring and leading to dull conversations about potential truths or answers to higher-than-thou questions. This existential approach is not taken in this book. Both Stoicism and Taoism can be used immediately and applied to your team from day one.

Stoicism is built on a practical application to solving issues you encounter daily, not discussing higher-level topics. Kings would use this philosophy to help deal with problems throughout their land. As a project manager, you are the "king" of your team. You must rule as such. The same applies to Taoism. It is a mindset shift. While you are the "king," you must think of yourself as an equal or lower among your team. In this way, your message will be clearer and more accepted, because you are speaking to them on an even playing field rather than talking down from your mountaintop. Each chapter in this book will have a discussion of both philosophies and how they pertain to the topic at hand. Again, you have the technical skills. Now it is time to add the exponential skills that will pay dividends.

Being different is another aspect of selling yourself as an entrepreneurial project manager. If people are starting to talk about how you have changed, you are doing something right. If you continue to go unnoticed, you will not have the influence on teams or senior management to get done the things you have in mind. Make yourself known for being the person who delivers. Use philosophy or motivation to get your team in the right mindset to take on the tasks at hand. Authenticity throughout is important while developing a new mindset. The way you deliver a message has more impact than the message itself. In grade school, there was a teacher who made this point through heavy-metal music. He said the lyrics would be about love or a relationship or a state of mind. Put to that style of music, it would always get his adrenaline going and the blood pumping. If you deliver a message about teamwork yet you are constantly in your office with the door closed, your message is not going to affect the team in the ways you had hoped, just like those songs lyrics. If they are set to a harp or piano, the message may be clearer and understood and gave the emotional response expected.

This book wants to bring out the child in each person who reads it. Using your child mind lends to the creativity needed to solve problems. Think back to when you were a kid and could play out an entire movie using blankets for capes and sticks for weapons. Now, as adults, those would seem silly and keep us engaged for a fraction of the time our child minds were occupied. When having team meetings, allow the kid to come out within yourself and others. You will be shocked at how many suggestions get brought up and the walls that come down with the simple change of mindset. Use that mindset to choose projects about which you have passion.

When I was in grade school, I wanted very badly to become a cartoonist. I would draw cartoons with different characters and try to create storylines— taking ideas from newspapers and reusing them, but with the people I had created. My mediocre talents were proudly displayed on the kitchen refrigerator and were read by an audience of two. Now, getting the chance to be creative is something I look for in a project. This book is one aspect of that creativity and why I wanted to write it. Think back to what interested you as a child and find those attributes in the projects your organization's executives select. Most of you may not have that choice, but you can find the attributes in the projects selected for you. People tend to find the negatives in something. I am one of those individuals. When you flip that around and look for positives, there are more than first recognized, especially through a child's eyes. There was a show on television in which people were shown a painting and had to name as many things that painting could represent. The number of items named generally ranged from three to seven. Once the presenter told them to see this painting through the eyes of a child, those numbers more than doubled. Our adult brains have limited our capacity to be creative through the experiences we have had.

When a project is similar, analogous estimating is useful. That same thought process cannot be used when trying to figure out problems for a new project. Because a project is similar does not mean the same techniques will work for it again. Road construction is a prime example. Roads all seem fairly similar on the surface. Below the surface is where the issues lie. Soil types can change from street block to street block. Utilities on a project can begin or end, depending on location.

Find projects or organizations that relate to your passions. There are projects out there that need your creativity to solve the issues. The result of the project can be creative too. Think of any building project. The nooks and crannies of that building are designed by someone with a creative mind. I had a boss who built his house using 45-degree angles for all interior walls. Most interior walls are at 90 degrees, but he decided to add that extra flare to turn an ordinary wall into a conversation. Even if you are not interested, you still have to ask, "Why?" Those little details of a project may seem mundane at the outset but can have impactful results on the customer or client. Out of 100 homes, how many have 45-degree-angled interior walls? The number is closer to 0 than it is to 100. If you are into sports, why not try to work on the project for building a new sports arena? How cool would it be to see your favorite sports team in an arena you helped build? Your passion may be writing. Think about how neat it would be to work with an author or director you have admired. These opportunities are possible.

Entrepreneurs rely on being a one-stop shop for the people doing business with them. This attitude is a huge asset and applies directly to project managers. How many times are you approached with a question and expected to answer immediately? No matter the subject, you are looked to for answers. This situation is where your entrepreneurial spirit is summoned. Entrepreneurs are looked to for ideas on solving problems they have not necessarily seen before. It is that creative mind showing itself. As a project manager, you will have to "fake it before you make it." Even when you think you have made it, there will be instances of fakery to portray the strong leader that you are. Do not take this as a weakness in your game. Show your team the spontaneous problem-solving skills it takes to become the best project manager. You are the chief executive officer (CEO) of your team, as is the entrepreneur the CEO of a business.

1.1. Philosophy

In college, philosophy was a required class to fulfill the graduation requirements. At the time, I treated it as such. Turn in the assignments with an attitude of "C's get degrees." Looking back, I should have taken advantage of the professor's knowledge of the area and tried to apply it to my life. My appreciation for philosophy has grown immensely since that time. Beginning to read more jump-started this involvement with philosophy. Instead of relying on people around me to structure my way of thinking, I began to seek out information in books. Books are a timeless resource that provides a glimpse into the trials and tribulations of others who have continued down similar paths in life. My approach to reading had always been that movies are so much better, quicker, and easier to hold my attention than books. A movie takes two hours to watch, while a book may take me two weeks to read. Of course, I would rather watch the movie instead. And, of course, people would always tell me the book is better than the movie. I would respond with an eye roll and continue about my day. Once work became slow and movies were not an option, I decided on books and have not looked back since. They have provided a wealth of knowledge from people I will never get the opportunity to meet. Philosophy books made me realize that these issues we have with projects and people have been around for centuries. There is not one right answer, but philosophy provides the mental tools to handle most situations.

I apply these teachings daily as a way to test myself physically and mentally. I go into the work day knowing that everything will not turn out the way I had hoped or planned. The idea is to see the problem through the eyes of the other individual as opposed to seeing only your own personal viewpoint. Take the time to understand the situation wholly and then come to a conclusion. I always allow myself to take a minute when something does not go right, to collect my thoughts and even move physically to arrange them in my head. Simply going for a walk or stretching for a minute can improve decision making immensely. Desk jobs constrict the physical movement one can do throughout the day. The thinking is that if you are not at your desk, you must not be working. Find ways to move around and keep the blood flowing. Do not sit stagnant because that is how you think the job should be performed. Perform the work the way you want, and the results will show. If you do not take this route, your life and career will plateau. Even improving yourself 1% daily, over time, you will see results unimaginable at your current state. The following paragraphs are historical descriptions of both of my core philosophies, to give you an idea of where each can be applied and why I choose them.

Stoicism comes from the root word "stoa," or porch, where the founder Zeno taught while speaking in Athens. It is built on the idea of the world being connected in a rational way and controlled by what the Stoics call "logos." Logos does not apply only on a universal level but also within the individual. The belief held is that logos has a plan for you and your role in the universe. You act "freely," but the decisions you make are made for you by the logos. As described in *Meditations* translated by Gregory Hays, "Man is like a dog tied to a moving wagon. If the dog refuses to run along with the wagon he will be dragged by it, yet the choice remains his; to run or be dragged" (Aurelius and Hays, 2002). As you see, the outcome remains similar, but the choice is up to the dog. The wagon is the logos. It is inherently good and should be trusted with its decision making. I compare it to a plan created by a project manager. The plan of the project is inherently good. It is scheduled to finish on time and within budget. Even if choices are made along the way, the plan is going to move toward to the ultimate goal of project completion. We want our teams to work as one, perform the standard operating procedures, and reach outcomes that are designed for project success. Also, project management problems occur with similar themes. In construction, the themes may be labor, maintenance, resources, scheduling, etc. The ways of solving these issues far surpass the issues themselves. Stoicism plays on those themes and tries to give you frameworks for attacking those issues. My take on philosophy is not one of some mystic voodoo and nuanced sayings to inspire people. It is to be practical and put to use. Change your mindset to reach levels unthinkable. Not to argue rhetoric, but to take it as it is. In golf, the saying goes, "Play it as it lies." That is my approach to using philosophy among project management principles.

The Stoic system talks about unity, order, and design. If that does not fall right into the knowledge areas of project management, then I am unsure what does. Reacting immediately to an issue with outrage or anxiety will lead to your personal detriment as well as the team's downfall. I have found Stoicism to improve my decision making based on the mentality it brings. You will run into people who make things difficult. We do not know why they act this way, but they will continually make your life difficult. Stoicism will provide the mindset that this person is an obstacle in your path to success. Through this person, success will be realized. It is your job to determine how this obstacle will be handled. Find out this person's idiosyncrasies and turn your enemy into a friend. More than likely, they want to be heard. Give them a voice in nonconsequential tasks to keep them occupied while you get things done. An entrepreneur encounters people every day who want what they have or are in the space the entrepreneur occupies. These are obstacles. You must find a way around or through them to get to where you want to be. No one said this was going to be easy, but the struggle is worth the gain. If you are a project manager, a promotion may be the goal. If you are an entrepreneur, a sale may be the goal. Either way, once the obstacle is overcome, both of those goals will be realized.

Taoism has a different approach than Stoicism because it is built more on the foundation of religion and mysticism. In my use of Taoist beliefs, I remove religion from the conversation to create a common ground for discussion. In this book, I will be referring most often to the *Te-Tao Ching,* which was written by a contemporary of Confucius named Lao Tzu. Its goal was to address the man who would be king and rule under the Taoist way. A project manager can be considered "king" of his or her team. People are looking to you for answers, direction, motivation, leadership, and so on. An analogy to use while describing the Way (which is one translation of the word *Tao*), mentioned throughout the *Te-Tao Ching,* is similar to a duck moving across the calm pond waters. On the exterior, everything appears to be at rest, as if the duck is floating, and the waters are as they are. Underneath that surface, that is the 10,000 things the Taoist philosophy discusses. The duck's webbed feet are moving frantically while the waters house fantastic life that is born again as the seasons change. Lilly pads and algae appear out of thin air, but meanwhile, the Way of the water has created this life. The feet of the duck have created the movement even though on the surface it appears as a smooth sailing duck using the wind to its advantage. This duck represents the Way of Taoism. It has the yin-yang effect. While it is calm, it is fierce. While it is gentle, it is harsh. Think of the undercurrent of the water and how it can sweep away everything in its path while remaining still on the surface. This attitude is something a project manager should approach each project, meeting, decision, and so on. Be the calm in the room yet strike with decisiveness.

Having duality requires a difficult balance. Use the Way to help enforce your management style. One who is bearing down on team members gets tiresome, and one who is uplifting grows old. There has to be that combination of breaking down and building up, and the Way stresses that duality perfectly. The Taoist philosophy wants people to return to some things more fully possessed by children—genuineness, sincerity, and spontaneity. Later in this book, the child's mind will play a key role in creativity and becoming solutions-oriented. As we grow older, learned behaviors tend to stifle our imagination and creativity. The Way wants our minds to reset and use some of those abilities we may have lost.

1.2. Sales

With philosophy, we know how to handle people in any given circumstance. Now, with sales, we are going to learn how to get what we want from those individuals. Sales is often a skill left to salespeople. Incorporating a sales mindset into your arsenal will take you to the next level. Imagine having the ability to communicate your ideas effectively not only to your team but also to senior management or executives alike. Taking your idea, project it onto a wall in front of a group, and show them how it will impact their immediate futures as well as the future of the organization. Bring your life to your concept. That is the idea behind sales in this book. It is not to take an existing product and go door-to-door with it competing for the trip to Disney World. It is about displaying your ability to communicate your ideas and "sell" your team and management on them.

I mentioned *the next level* above, but what does that mean? In every year-end meeting, the senior manager stands at the front of the room stating that the team needs to take it to the next level, and every year it is the same message. If you are projecting the same message year after year and not getting results, that is the definition of insanity. The "next level" concept must be concrete in its actionable steps. Define the next level for your team. If it is 10% increase in sales and a 5% increase in profits, tell them your goals. Do not leave the ambiguous term "the next level" hanging out in the ether and hope they can figure out what that is along the way. Managers can fall into the trap of hearing concepts, telling people about them, but never creating actionable steps to achieve those results. Give them direct milestones to reach. Exaggerate the numbers to give them truly next-level results. If the goals are too mundane, there is no incentive to reach them. The concept of the next level sounds good in meetings in front of your team, but to continue to use that term with no results will cause confusion and eventual mistrust among the team. I will explore the details later in the book about setting goals and helping your teams achieve them.

In grade school, every year there would be a magazine sale. We would be handed a brochure on all the magazines available and try to sell as many of them as possible. At certain stages of selling, wristbands with various characters on them were awarded. I recall looking at those cotton balls with glitter on them as if they were from the moon. I wanted each of them badly. I wanted to sell one hundred subscriptions to everyone on the planet so I could get the craziest wristband and wear it proudly around the school. At the end of the competition, the grand prize was a limousine ride to school with ten of your friends and a pizza party at lunch. Of course, every year I would end up selling the same ten magazines to the same four people and reach only the first stage of prizes. I would look with envy as kids would accumulate five, six, seven cotton-ball items dangling from their wrists. Looking at them and thinking how in the world that was possible. Being a young, naïve third grader, I did not realize that those kids had parents who worked at the front desk of large corporations, who would sell the magazines for them. I thought they knew something I did not. My Mom would purchase three or four magazines, and that was a huge sale. *Sports Illustrated for Kids* ended up being the best thing to come of this entire competition. Since then, I sold numerous other items for fundraising events, but never once did I think of selling my ideas. To me, selling always involved a physical item. An exchange of goods or a zero-sum game if you will. I give you one thing, and I receive one thing in return. The focus of selling in this book is eliminating that zero-sum game mentality and building on each others' ideas and concepts—using each other as a springboard.

One source of learning I never thought possible is learning from my dog, Belle. She is a rescue dog advertised as a lab/shepherd mix. After giving her a DNA test, it turns out she is a mix of every dog imaginable. They labeled her a "mega-mutt," which has odds of 100,000:1. With that background, how does one go about training her? Usually, breeds have a temperament and training style best suited for them. Having a smidge of everything, no single attack plan works for her. This approach has proven difficult. Taking her to the dog park was working until she found out that small dogs are her favorite toy. Going to doggy day care was an option until she came home with a mysterious gouge in her back. Using a training collar has proven effective (vibration setting only), yet the second you take it off, she is back to her normal behavior. Does this resemble a team you are trying to get to perform? No matter the angle you attack from, there does not seem to be a way to get them to coexist. Techniques of approach to the dog and your team have similarities that will be discussed throughout the chapters of this book. Some may seem far-fetched (pun intended), but this book is trying to make connections where there do not seem to be any. Open up the creative brain. Let your imagination run wild, and the connections will be there.

In listening to a webinar on strategy and innovation, Robbie Bach, who is a former president of the Entertainment and Devices Division at Microsoft, coined the term *intrapreneurship*. The spirit of intrapreneurship is the idea behind this book. Intrapreneurship is taking your entrepreneurial mindset and applying it within your organization. Entrepreneurship is not about starting your own business; it is about making something extraordinary about yourself. Using the tools and techniques you have acquired through the years of experience and turning that into a leadership role that inspires others. Taking risks and expanding innovation are at the heart of entrepreneurship as well. Within an organization or project, calculated risks need to be taken for growth. Innovation is at the forefront of what we do every day as project managers. Using the best equipment and having the best people as part of our teams sparks innovation.

Rewards for taking on those risks should match or exceed the potential consequences. A pat on the back for putting your neck on the line does not equate. What would be the motivation moving forward to take on big challenges if the reward does not fit the risk? You must reward those willing to seek risk, put their necks on the line, and find success with all of those roadblocks in the way. An entrepreneur will be rewarded with fortune relating to the business created. Project managers must be willing to do the same for the team members putting themselves out there the most. A company I worked for based their bonuses on a company-wide "pool" rather than on individual levels of performance. Where is the motivation to work longer hours and take on bigger projects if the reward is based on the success of the company as a whole? If the company struggles even though your performance shined, your bonus is cut for that year. If you do the same work this year that you did last year, and the company thrives, your bonus goes up. It is counterintuitive to reward individuals based on an organization's efforts. The only real control an employee has is performance, day in and day out. With that type of reward system, you are taking away the one thing an employee can control and be objectively judged on.

Taking this new approach is going to take a commitment from you. If you do not believe in your ideas, who else will? Why would anyone else care if the person in charge is not fully committed? Take the "fake it until you make it" approach if necessary. These new mindset shifts will change the way your team operates for the better. Committing yourself to them will inspire your team to do the same. Think about a situation where all you needed was for someone to encourage you to do something uncomfortable. New approaches will be uncomfortable at first. It will be like staring into the deep end of the pool when learning to swim. The easy parts are over. There are no more feet touching the bottom to save you. There is no more life jacket to keep you afloat. It is all up to you. Sink or swim. These ideas and mindsets will sink or swim depending on the effort you apply. Continue to build trust with your team through results and gain further commitment.

1.3. Imposter Syndrome

While taking on this new approach to project management, you may come across the *imposter syndrome*. The imposter syndrome is the feeling that you do not belong in the position you are in because you lack something. That something may be experience, a degree, a credential, whatever. However, results are what matter. You could have a background in education yet run a successful project in construction. If you make the organization money, do you think they care that your degree was in education? Of course not. Take that approach to implementing this book in your practices. It will be new and uncomfortable, but if you stick with it, the results will be there. Remember your first day on the job. You meet a bunch of people. Maybe you remember one or two of their names, and then you go right to work. Your boss already has three or four tasks for you to get done, and you do not even know where the bathrooms or cafeteria are. How uncomfortable was that? Now look at you. You know every nook and cranny of the office. You are giving people tours on their first day. My, how times have changed. You will take these techniques and mindsets, make them your own, and before you know it, it will be old hat for you. Your team will be performing at an all-time high with greater morale. You will be exploring ways to create more revenue rather than putting out every fire. Your team will be empowered, allowing you to focus on high-level goals. Your impact on your team will not only be recognized but also your impact throughout your organization. Your ways will rub off on the others and get noticed by senior managers. Before you know it, you will be in that class.

Throughout this exploration of entrepreneurship in congruence with project management, see how your mindset shifts. I want you thinking in new ways and attacking problems from a different perspective. Because you learned something one way does not mean that is the only way. Because you did something in the past does not mean the future has to go through that same experience. You already have the tools. Now it is time to explore how to use them. What good is a hammer if you do not swing it? What good is a schedule if you do not know how to navigate it? My goal is for you to become creative far surpassing the current model for what a project manager does.

In the next chapter, the traditional organizational structure will be discussed. What are the positives and negatives along with benefits of keeping things as they are? What ways can these "best practices" be improved? What can you do as a project manager to ensure that not only are best practices being used, they are also being revised? These questions will be answered and much more in Chapter 2.

Chapter 2

Traditional Mindset

If the task takes longer to delegate than it would to perform, *do it yourself!* There are numerous times as an assistant project manager when the task I am being asked to perform takes longer to explain than it would if the project manager had decided to do it. In compiling a quote for work, there would be general information required such as the name of the contact person, company name, location, and date. This information would be given to the project manager after the quote was compiled. It is not required for the estimator to put the quote together because the quantities are known. Once the quote is complete, instead of putting that information in the system, the project manager decides to delegate it as a form of acting as a project manager. Never act like a project manager! Be one, and delegate tasks of relevance. If you continue to delegate remedial tasks, your team expects lower performance from themselves because these tasks mean so little on the grand stage. The quoting system is not kept local, so each team member, including the project manager, has access to the quote for data entry. Laziness comes to mind when I have been assigned a similar task. If the project manager cannot input the information, what other tasks is the project manager unwilling to do? It sounds like a nitpicky characteristic, but these things can become major issues. By the time the project manager reads the information aloud, you write it down, and then enter the information into the quote, the project manager could have easily entered that information within the same time frame. Keep this in mind when delegating tasks to team members. Make sure they are relevant to project success and require work to be performed. People do not want to be relegated to data entry for the sake of data entry.

The traditional way of scouting player talent has come under fire in recent times. People are moving toward a mathematical approach in trying to determine who would fit best within their organization. The National Football League Combine has long been the measuring stick for football talent. They use measurements such as height, weight, wingspan, hand size, 40-yard-dash times, broad jumps, and so on, to determine how a player will translate to the next level. This approach has delivered massive downfalls, yet it continues to be the preferred way of deciding who can play and who cannot. The "eye test" has been another method for determining talent. Instead of looking at the metrics of an individual, people will study the game tape and look at their performance in game settings to predict future performance. Think of ways your organization may be using such tests to predict future employee performance. Are these techniques working? Should they be adjusted? What adjustments would you recommend?

Human resource management is an understudied, overlooked aspect of the project management field. Human resources are your most valuable resources, bar none. Finding out where top talent resides, what characteristics breed top performers, and what expectations to set for your organization are secrets most company executives tend to overlook. Take surveys of your top performers. Find out what makes them tick, what they look for in a position, how they use their time, and so on. Use that information to seek out similarly linked individuals.

The eye test is often used in the construction industry. People may talk a big game, but when the rubber meets the road, they cannot perform as expected. I use a sports analogy to better understand this concept. Say that a football player performs at a high level in college. Because of this performance, the player is considered a top prospect for the professional level. Before the player is drafted, however, he is run through a gamut of physical tests. If he performs poorly in these tests, his draft stock plummets. Instead of relying just on the information from actual game play to determine value, scouts look at these test numbers and come to a different conclusion. The opposite is also true. A player can test well and improve his or her draft stock without the game play numbers to back that up. Which "test" has greater impact on your decision making, the one in which the player performs well under pressure in game situations, or the test where the players has months to prepare for specific, known drills? I prefer the "eye test," seeing how the player performs under stress in an actual game. I want team members who have put in the work daily, not just ramping up for a big event but then falling back to their baseline performance.

People want to share information. Allow them an avenue to tell you freely how they do things and what characteristics they possess that makes them who they are. One survey can change the way you search for future employees, how you hire them, and even how to train current employees to mimic the traits of your best. Ask meaningful questions. Going back to the NFL Combine example, knowing what someone's favorite color or food is does not translate into how that person will perform. Even how fast a person runs does not determine how good a football player he will be. Take a look at the top 40-yard-dash times in recent NFL Combine history. Electronic timing, which is more accurate than hand stopwatch timing, began in 1999 and is the starting point for the recorded times as shown in Figure 2.1.

Name	40-yd Time (sec)	Years Played	Pro Bowls
Rondel Menendez	4.24	1	0
Chris Johnson*	4.24	9	3
Jerome Mathis	4.26	7	1
Dri Archer*	4.26	3	0
Stanford Routt	4.27	9	0
Marquise Goodwin*	4.27	4	0
Champ Bailey	4.28	16	12
Jacoby Ford	4.28	6	0
J.J. Nelson*	4.28	2	0
DeMarcus Van Dyke	4.28	6	0
Fabian Washington	4.29	7	0
Dominique Rodgers-Cromartie*	4.29	9	2
Josh Robinson	4.29	1	0
Darrent Williams	4.30	2	0
Tye Hill	4.30	5	0
Yamon Figurs	4.30	5	0
Darrius Heyward-Bey*	4.30	8	0
*Currently on 2016 NFL roster.			

Figure 2.1 NFL players' times for the 40-yard dash. (Information courtesy of NFL.com.)

As you can see from Figure 2.1, being the fastest does not mean being the best. Scouts are enthralled with a player's ability to run fast and forget the other facets of the game. Out of 100 seasons played by these players, only 18 resulted in a Pro Bowl season. Of those 18 Pro Bowl appearances, 12 belonged to Champ Bailey, who will be inducted into the Hall of Fame when his opportunity arrives. Jerry Rice, one of the game's best receivers, was not the fastest or the strongest or the biggest. It was his mindset and his attitude toward details and precision that gave him the ability to achieve greatness. Find these same attributes among your team members. Once you have found similar traits, go out and find people with those traits to acquire for your team. Conventional thinking leads to conventional results. If things are not going well, change your ways and realize the results.

Traditional mindsets lead to statements such as "I want all documents printed single-sided." When you are in an organization that receives revenues of billions of dollars annually, think of the paper wasted in that request. People are constantly looking for money-saving ideas when many of them may be right under their noses. While the money is not coming directly out of your pocket, there is a budget somewhere that is getting destroyed because you have a preference that can easily be avoided. Printing thousands of pages when only hundreds are needed not only impacts bottom lines but our environment as well. After becoming aware of the benefits and yet continuing to plod on is a level of arrogance unreached by many. Having a preference is not the issue. The issue is realizing the benefits of change yet continuing on your path.

Assumptions can be categorized under the traditional mindsets organizations that have. They have developed a road to success, and swaying from that path does not interest them. The relationships developed around assumptions may have been the key to success in the early days. "If it ain't broke, don't fix it." An entrepreneurial project manager does not stand for that as a reason. You need more. You expect more. Challenge assumptions. Use synonyms to develop a different way of looking at the assumption. Even the word *assumption:* Replace it with *belief, expectation,* or *speculation.* Those words lead to a different thought process than an assumption does. If you consider expectations to be the same, what does that mean throughout your organization? Does it mean the same thing to different people?

An employee assumes a holiday bonus because that is the tradition the organization has instilled. Why not flip the script? Why not give employees a bonus *before* the busy season, as an incentive to work hard. Reward, then reap the rewards, rather than reap, then possibly reward. Harry Seifert, CEO of Winter Garden Salads, did just that. Before the busy season, Harry gave each of his employees a bonus. As a result, production increased 50% (Michalko, 2006). Why is that so? Think of a time when you worked more hours than ever. The office felt like your primary address. But then the holidays came, and your bonus check was lower than the year before. Why? The answer is because that is how things shook out this year. Thanks for your effort. Try again next year. How does that make you feel? Do you want to work even harder the next year when your bonus may be even lower? If you were given a bonus before working all of those hours, would your actions have been diferent? If so, why? As an individual who wants as many knowns in his life as possible, I would love to receive my bonus before the busy season.

2.1. The Midas Touch

Most people are aware of the story of King Midas and his ability to make everything he touched turn to gold. Does that sound comparable to a manager or a team member you have met? Countless people give themselves far too much credit. It is one of my pet peeves to hear someone talk about how, without them, there would not be (fill in the blank). After leaving a large corporation, I knew full well that they would replace me without skipping a beat. How could I be so confident as to say that without me there would not be a success? Of course, projects I had been involved with were successful, but there were also projects that were not successful. Having an attitude that everything I am involved with will be successful can create a false foundation, and things can easily crash. Issues can become overlooked because of past successes. You will hear throughout this book about not allowing past success to predict future results. Every project is different, new, and stands alone. There may be analogous elements, but to transfer past successes to future projects will create unforeseen issues that were not accounted for because they did not happen on the previous project. Traditional companies with a top-down management structure lean toward senior managers having the Midas touch. What they say goes. Not to challenge that approach is dangerous for the future success of the team and the organization. A constant check and balance are necessary so that neither party gets skewed and can stay on course.

At my most recent organization, money was not an issue. It was an expendable resource. If you needed something, buy it. If it was tens of thousands of dollars, justify it reasonably, and it was yours. There was not a big concern. It cost the company creativity in solving problems. There was not a need for root-cause diagnosis to figure out the issue. They would buy a tool or piece of equipment to fix one issue, which would lead to another issue for which they would buy a tool or a piece of equipment to fix that issue, and so on. Rarely there was an effort to get to the heart of the issue. For them, more management always seemed to be the answer. If one project manager could not solve the issue, put a second manager on the problem. With this approach, problems continued to manifest themselves. No real solution was ever brought to the table. It was a bunch of quick fixes that could lead to more problems. Instead of controlling the fire and finding a way to douse the flames, they continued to throw buckets of water, thinking that would do the trick. Take a step back and analyze the situation. Get to the heart of the cause and spend the endless resource wisely.

It is easier to write checks than it is to put in the tough mental effort that problems require.

For a company that had endless resources, they did make that money by introducing contradicting approaches to the industry. It was involved in so many different aspects of the industry that if one were down, one would increase. The stock market is an analogy. If interest rates are down, bonds are up, and vice versa. Keeping that approach to your team and the projects you are working on will reduce the risk of taking on more. If you balance the portfolio correctly, there will never be a down time. Putting all your eggs into one or two large projects and creating a boom-and-bust atmosphere runs the risk of ultimate failure. Planning for failure becomes a way for success.

Contingency reserves are built into every project for a reason. You know things will go wrong; it is inevitable. Lower the odds of failure by diversifying projects, teams, and individuals. In the book, *What I Learned Losing a Million Dollars* (Paul and Moynihan, 2013), the authors say that learning *how not to lose money* is more important than learning *how to make money*. Often, that concept is overlooked. Organizational leaders are looking for ways to make money. They often tend to take on more projects, or they may create more divisions within the company. Instead, take the approach of *How do we save money?* Is the new project worth doing, or is it just to keep the employees busy? Do we need to get involved with those other aspects of business, or should we hone our skills with the divisions we already have? The goal is not to make each division perfect before adding more. It is to instill this attitude among that division so the staff members know to keep costs low and look for ways to bid projects where savings can be had. In construction, trucking is a massive cost, whether it is hauling material off site or bringing material to the job site. An industry-standard way to save on trucking is to find a site on the job where the material can be stockpiled, rather than having it off-site. The off-site locations will be as close to the project as possible, but on the site is always best. Working with the inspectors and owners of the project to keep the most material on the site is a way to take your bid prices and reduce the cost. At $100/hour for a truck, if you can save a round-trip time of 15 minutes, that is $25/hour you are saving by keeping the material on the site.

2.2. Setting Goals Through Stoicism

> *If you don't have a consistent goal in life, you can't live it in a consistent way.*
>
> – Marcus Aurelius, *Meditations*

Companies need employees to have a goal throughout their tenure. During the job interview, the classic question is, "Where do you see yourself in five years?" This question gets right to the heart of the issue of having a goal. As Marcus Aurelius pointed out two millennia ago, if you do not have a goal, you cannot work toward that goal. The road becomes much clearer if you have a destination in mind. The path will not be straightforward. Do not confuse the path with the goal. There will be obstacles (as we will discuss later) to deter us from our goals, but that is no excuse for abandoning the mission. Keeping your goals aligned with the organization's goals leads to consistent performance and becoming a valued asset to the organization. Remaining consistent allows everyone around you to get a feel for how you work and helps them work in tandem with you. Consistency lends itself to fairness. While life is not always fair, remaining consistent in how you deal with issues garners respect from your team members and senior executives alike. Remove bias from decisions, whether it is project-related or personal. Handling people properly will make your path to your goal much more accessible and attainable. The goal of any project is to meet requirements successfully, on time and within budget. Maintaining consistent performance, both individually and as a team, will lead to meeting your goals more consistently. Therefore, have a plan and follow it.

2.3. Taoism

With the highest kind of rulers, those below simply know they exist. With those one step down—they love and praise them. With those one further step down—they fear them. And with those at the bottom—they ridicule and insult them.

– Lao Tzu, *Te-Tao Ching*

Apply this quote to your organizational chart and determine your place on the chart. Are these the attitudes you find among employees under your oversight? Be cognizant of the fact that you will not please everyone all the time. Know that there are times when you have to upset people, whether it is to motivate or to get the point across. As a project manager, you are the "ruler" of your project. When you are dealing with people, recognize that some people are aware only that you exist—and nothing else. They do not have a personal or professional history with you. You are the contact person for the project, and that is it. Others on the project, mostly team members directly below you, will "love and praise" you because you are their manager. You have influence over their immediate futures. Beware of having "yes men" around. Conflict is an essential part of a successful team. Without it, new ideas and innovation are stifled. Below that level of employee, people will fear you because you are their manager. They do not have the knowledge to be confident in their beliefs and fear your intelligence. Remember when you were the newcomer or an assistant project manager and consider how you took direction from the project manager.

Did you fear situations when it was your turn to speak? While you have the ability, the knowledge is what keeps you from fully developing a voice. This struggle is the state of fear at that level. It is not the intense fear of losing your job, but the fear of looking stupid when you are asked to speak. It is your job to delegate tasks that build this confidence to the point of love and praise. The people who ridicule and insult you are the ones you have delegated to do the individual tasks themselves. In construction, it would be the laborer or the operator. They do not see the daily planning it takes to get them to this point. Their focus is the intense heat of a summer day coupled with the long hours needed to get the work done. Their focus will not change. Whatever the explanations you give them about what went wrong and how you are at the mercy of the sponsor, it is still your fault that they are doing the rework. Taking the Taoist method and noticing each level of employee will give you a better understanding as to how to approach each person when discussing possible ideas. The level below will love the idea. The level below them will be in fear and therefore agree to your idea. The level below them will hate your idea. Find employees who love and praise your work enough to offer up counterpoints that will help to improve the idea rather than sabotage it.

2.4. Sales

When discussing selling, people picture a well-dressed man with slicked-back hair trying to buy low and sell high. Traditionally, organizations have the sales staff to push their products to customers, create buy-in for future purchases, and help to establish the brand for the organization. Their goal is about closing the deal and then moving on to the next opportunity. Increasing revenues result in higher commissions for the sales staff, driving them toward more sales. This idea of sales is isolated and creates separation between operations and sales. While it is important to have individuals who are skilled in sales, it is equally important to apply some aspects of selling among everyone in the organization. Keeping sales isolated leaves salespeople out of the loop as to how the product is manufactured and the true essence of what the product delivers. Anyone can read specifications and measurements, but it takes a true salesperson to make the connection from the manufacturer to the customer.

Sales are an often overlooked skill. The attitude can be that if customers want the product, they will buy it. The idea behind becoming effective in sales is to make customers realize they need the product. Basing sales on price alone is a costly error that is commonly made. If the item is inexpensive, the idea is that people will buy more. This happens in construction especially, because projects typically are awarded to the lowest bidder. The service an organization provides should be weighed along with price and location. A former employer of mine would always be in the middle of the pack on price yet continue to get the work. They offered the owner a service that others could not. We were so established in the industry that people knew we produced a great product that would pass inspection without hesitation. The peace of mind we created for our customers was worth the extra few cents per unit. The professionalism with which we approached each project was second to none. Constant communication with the customer was "selling" the idea that we are experienced in this process, and you have nothing to worry about because we will deliver successfully. Following up with customers before, during, and after the sale can make the difference between getting the job or not. While our price was not the lowest, the mentality we brought with it landed us the projects.

For future projects, the organization would be the only bidder. Bidding against ourselves from previous years would be the web we wove. How much should we increase from previous projects? If there was a year in between, we would have a percentage. If the project were to come later in the year, a different percentage would be applied. The organization would increase wages in June, so the price would go up after that date. It was a tricky balance between covering costs and trying to maintain a healthy relationship with other organizations. Since we were the only ones bidding on the projects, we did not want to price-gouge because the sponsor could consider other companies if we became too greedy. Positioning in the marketplace was important. Being the sole bidder on projects remained a priority. Holding that position was the key to success.

2.5. Belle the Dog

Belle is an energetic (to say the least) puppy who is constantly on guard and responds to every sound and movement outdoors. Therefore, puppy class was a necessity. The traditional methods of getting your dog's attention include treats, making noises, and calling to your dog. She responded to these methods with an attention span long enough to get the treat, and then proceeded to check out what was going on around her. Dogs of all sizes attended the class. She was most concerned with the smaller dogs, constantly trying to pull me toward them so she could get a closer look. The traditional methods of puppy training were not going to work for this special breed. Eventually, I resorted to giving her medication to focus her energy, and she became a great dog.

Apply this to a team member of yours. Do you have a team member, no matter what you do, who does not take direction? Similar to Belle, you have to seek alternatives instead of giving up or resorting to threats. Whether the techniques are aversive or reward-based, use them until you find what works. Withholding something from a team member may work best. This technique is like the seat belt buzzer in your car. When you put your seat belt on, the buzzer turns off. You are being trained to put your seat belt on because you do not like the buzzer constantly sounding. Continually calling someone to check up on them to get a task done is another example. Once the team member starts to perform proactively, the phone calls can lessen. No one likes to receive phone calls throughout the day. Usually, they signify problems, so when a team member can lower the number of calls, that is an aversive technique.

Reward is a more common technique to get results. In the dog's case, treats were her reward for performing the desired action. In your team member's case, a bonus is an example. Most people work because they get paid. Paying them more for desired actions becomes a way to increase the volume of those actions. Lay out criteria for your team members, such as once they hit a level of performance, they are rewarded. The reward does not always have to be money. A free lunch or T-shirt will do the trick and is more cost-effective. Reward techniques have worked best for Belle. She responds better when there is a prize readily available as opposed to the aversive techniques. While your team members are not dogs, the framework of the techniques remains very similar. How you go about approaching your team members is up to you. If you are lucky, their jobs and paychecks will be rewarding and satisfying, so additional techniques will not be necessary.

In construction, I find workers doing things because the people who came before them did that task that way. In management, it should be your goal to find the most effective way to do things, not the way you necessarily did things. If you shoveled off conveyors and now a skid steer can do that same task, have the skid steer perform the work. If you lifted heavy objects and there is a forklift available, have the machine perform the work. I see many instances of people lifting heavy objects, and there is a machine sitting right next to them that could not only lift the object more easily and safely but would also save the energy of the worker for more important tasks. I hear stories from people, looking back fondly, about those times of hard work. The only reason they are mentioned with such fondness is that they do not have to do them anymore. If you made it out of the struggle, you look at the bad times in a different light than if you were still dealing with the mess at hand. While shoveling off a conveyor, I do not think I have ever heard a laborer discussing the task with reference to it building character. The manager smiles because those days shoveling endlessly are over, and now the manager can delegate that task to someone else. Never make someone do something because you had to do it. I find this form of initiation frustrating and needless. Some traditions deserve to go away. Times have changed. Ways of doing things are different and vastly better.

"The appearance of busyness reinforces the perception of causality, of the link between results and one's role in them," wrote Nassim Nicholas Taleb in

Black Swan (Taleb, 2007). The construction industry provides classic examples of people being busy and thinking that busyness is directly causing work produced. I remember being a laborer, and people constantly telling me to stay busy. In my head, I am thinking there is nothing to do. Their perception is that there is always something to do. Because you look busy does not relate to progress. You may be a detriment to progress with your busyness. I wanted the tasks I worked on to impact progress positively, not just move around for the sake of moving around. Most laborers are comfortable with that action, moving to move instead of moving to produce. Managers drive around ensuring that people are moving, and if not, making them move—such as giving them tasks so they will appear busy while no true work is being completed. Can you think of anything worse than doing work purely for the movement, not the intent of progression? There is nothing more frustrating than to hear your boss shout at you about some task you know can wait, but your boss wants to see movement. The real issue is crashing the project with human resources. Maybe it is a slow time for the company, and managers do not want to lose good employees so they put people on a project where they are not needed so they can hang on to them for when they are required. If that is the case, all involved should have an understanding of the lack of work.

Operations managers have a difficult time with the lack of movement. In my experience, they appear the busiest of all. They seem to be constantly on the phone or running around to the next meeting. I wonder how much "work" they get done. It is similar to the physics equation of work. They can move around all they want, but if the object comes back to the original base point, no work has been done. To suggest this behavior would be sufficient to fire someone, one first needs to ask: How many phone calls are made in which nothing is said? How many miles are driven to where nothing is done? How many meetings are held where the plan changes minutes after it is over? These appearances of busyness cost your teams and organizations money. Do not correlate busyness with progress.

2.6. Importance of Trust

"An engaged workforce is a primary driver of results," says Stephen M. R. Covey in *The Speed of Trust* (Covey and Merrill, 2006). He argues that the way to build an engaged workforce is through trust. With that trust comes engagement of your team members. Imagine having to listen to someone you know lacks expertise. Would you want to work the extra hours and take on more projects for that person? Of course not. You would feel that you were being used for a cause, and it is not a cause you support. How would that impact your engagement on the projects? Negatively. "People do not quit jobs; they quit bosses." I have heard that sentiment many times, and to me, that statement could not ring more true. The trust you need to build as a project manager, as a leader of the project, is crucial to its success. Never take that for granted, and do not abuse it. Trust takes years to build, seconds to break, and forever to repair. Do not take trust lightly. If your team does not seem interested in the project, take an look inward at the message you are conveying. Do you believe what you are saying? Do you think this project will deliver what you are saying it will? If not, the trust your team has in you is suffering, and the project suffers too.

Trust within your team is important, but it becomes magnified on an organizational level. "Culture eats strategy for breakfast" is a familiar phrase used in project management. If the organization has a trust issue, the initiatives will fail. A company I worked for tried a huge initiative of trying to get each division to work together. Instead of bidding out the work to other companies, it wanted to self-perform as much work as possible. Because of the company's size and culture, the initiative never got off the ground. Monies were distributed unevenly and not according to work performed. Senior managers had trust issues with the C-suite executives making matters worse. How can an organization expect people to work together if, behind the scenes, people are selfish? It is impossible. Even if the corporate culture is uncontrollable, however, you can control your team's culture. Create a trusting environment that engages the team members. Consider their opinions. Allow for ideas to be bounced around rather than discarded disparagingly. You have the ability to influence your team; take the opportunity to do so. No matter the culture around you, use your local influence to change for the better. And it all starts with trust.

2.7. Low Friction vs. High Friction

Big projects bog down a portfolio. As an analogy, think of a tractor driving down the highway, and traffic backed up ten deep. Or when you eat a big meal and want to take a nap. Or a big log is added to the fire, dousing the flames. When your organization lands a big project with a long duration, resources tend to bog down and spread thin. You may have to promote people who are not ready for the challenges ahead. While a major project creates stability and lasting revenue, it may require the use of less skilled resources. Small, manageable, bite-sized projects consecutively are the ideal situation. Maximize your labor pool by keeping them busy and moving forward. One issue per project over five projects is easier to handle than five issues on one project. The cumulative effect of those five issues will be difficult to recover from, as opposed to resolving one issue and then moving on to the next one. Remember, a race that used to be won by the biggest will now be won by the fastest. Change is inevitable. If you have large projects bogging down a portfolio, the ability to change quickly is impacted negatively, resulting in falling behind in the rat race that is business.

A low-friction company requires little capital to realize greater market value, while a high-friction company uses the greater capital to see the higher market value. Examples are shown in Figure 2.2 (numbers reflect September 2016 evaluations).

Company Name	Assets	Market Value
Apple Inc.	$290.48	$636.41
Amazon.com	$65.44	$360.26
Exxon Mobil Corp.	$336.75	$345.36
Wal-Mart Stores Inc.	$199.58	$222.66

Figure 2.2   New York Stock Exchange, 9/15/16 (dollars in billions).

As you can tell from Figure 2.2, Apple and Amazon run far leaner operations than Exxon Mobil and Wal-Mart. While they are all wildly successful companies, their operations and strategies differ. Organizations are constantly aiming to produce more with less. Increasing technology helps in that effort. Maintaining a low-friction mentality allows for optimal fluidity in operations. You can move from one project or idea to the next with relative ease by running lean and keeping projects to a manageable size for your resources.

2.8. Passion Without Purpose

Find something you love to do, and you will never have to work a day in your life.

– Anonymous

How many times have you heard that quote? Or people close to you telling you to follow your passion? This traditional way of thinking has always bothered me, because I see people following their passions all the time yet living in a friend's basement or worse. That lifestyle does not appeal to me. It appears that following your passion does not lead to where I want to go. Ryan Holiday, the author of *Ego Is the Enemy* (Holiday, 2016), talks about passion needing a purpose to be successful. His example of dogs having passion struck a chord with me.

As I write, I am watching my dog Belle leap high atop a fence chasing squirrels running from tree to tree. She is obviously very passionate about these squirrels yet never seems to accomplish her goal of catching one. Same goes for flies. She loves chasing down flies yet cannot seem to capture one. Both activities are overflowing with passion. However, if her job were to catch squirrels and flies, she would be terminated rather quickly for her lack of execution. Belle's passion is a hobby, not a business. One day, she might catch a squirrel or a fly, but happenstance is not a successful business plan.

I like to relate passion without a purpose to a flame without oxygen. It will burn brightly, fascinating those around it, but it will not provide any of the effective characteristics of a flame such as warmth or ability to cook. Without a purpose, passion is useless. I do not recommend following your passion solely. There must be a reason to stick to it. Finding a passion is easy. Finding the purpose of the passion will take you from your friend's basement to the penthouse.

In Chapter 3, I will begin to discuss how to take an entrepreneurial approach toward your organization: how to take these traditional ideas about management and use that framework to your advantage. Knowing the lay of the land is a big advantage in getting what you want. Influencing the right people at the right times will give you an edge in getting the correct resources for your project. Using meetings effectively is an example. Instead of going through the motions, get to the point and ask the right questions to get meaningful answers.

2.9. Old School: Kicker and High Jumper

George Blanda, a former NFL football player, played for five teams from 1949 to 1975, an unprecedented 26 years as a professional football player. At the time

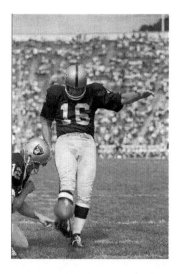

Figure 2.3   Hall-of-Famer George Blanda (Blanda, 2016).

of his retirement, he was the all-time leading scorer in NFL history. He played the position of place kicker and quarterback, another facet of old-school football you will not see today. Blanda is the only football player to play in four decades (the 1940s, 1950s, 1960s, and 1970s). If you watch old films of Blanda kicking field goals, he has a straight-on approach with a straight follow-through, as shown in Figure 2.3.

Blanda's approach to kicking field goals was literally straight forward. There was no fancy setup or routine. It was step straight back three yards, wait for the ball to be snapped, run, and kick the football. This approach was the traditional way a field goal was attempted in football until Pete Gogolak appeared in the league and would change the game forever. I will discuss Gogolak in Chapter 3.

Track and field meets have an event called the high jump. Athletes must clear a bar leaping off one foot with a running start. The athletes take turns until the greatest height is cleared by a single jumper. The technique for clearing the bar has changed dramatically over the years. Originally, athletes would take a straight-on approach, running directly at the bar and trying to jump over it. Next came the scissor approach, which was alternating your legs over the bar as you tried to clear it sideways, as shown in Figure 2.4.

Figure 2.4 Scissor approach to the high jump (Scissor Approach, 2016).

The scissor approach was effective but not efficient. There were greater heights to be reached with better techniques. Eastern cutoff, Western roll, and straddle were all techniques developed after the scissor technique, and all were more efficient in clearing the bar. Valeriy Brumel set a world record 7 ft 5¾ in. using the straddle technique. There was still more height to be had, and Dick Fosbury was the guy to get it. I will discuss his approach and accomplishments in Chapter 3.

I mention these techniques because they were effective and efficient for decades in their sports. Running straight at the football and kicking it through the goal posts worked. Running at an angle and jumping sideways while clearing the high bar worked. So why fix it? The question is not, "To fix or not to fix?" The question is, "How can we make this approach better, more accurate, and produce more predictable results?" Project managers have systems in place that work. They have worked for many years, like the old systems for kicking a football or clearing a high bar. Improving on those systems and not accepting them at face value is our aim.

2.10. Waste in Traditional Organizations

In a study carried out by Lawrence Cooper (Cooper, 2016), he found the primary waste in a traditionally run organization is that innovation gets killed. The red tape does not allow for easy transition or organizational agility. If you want change to occur, be ready to wait. You will have a meeting with your boss, who will have a meeting with senior leaders, who will have a meeting with a board, who will then vote and finally a decision is made. The string of jumps and hoops to get anything in motion is daunting. From conception to reality, it may take years. The latest technology is already obsolete. How can an organization afford to take months, if not years, to decide on a potential innovation? It cannot. With this traditional setup, innovation gets killed because of the twists and turns one must take to initiate the change, let alone realize the change.

Cooper also lists talent gets lost, devastates purpose, and process limits productivity as other primary wastes in a traditionally run organization. Do any of these wastes ring true? How would you rank them within your organization? Throughout the coming chapters, being talented, finding your purpose, and becoming the most productive project manager will be talked about extensively. These three categories are stunted in a traditionally run organization. I hear project managers talk about talent when it comes to important characteristics of a successful team. My stance is that talent is all around us. It is our job to find it. In a large organization, when the project teams are at capacity, it is difficult to step back and assess the talent you have, but it must be done. Not recognizing that talent leads to these team members losing their sense of purpose. Talented people want to contribute and demonstrate their skills to make the team and organization better.

2.11. Leadership Modeling

With the wastes identified, the organization will want to change. The entrepreneurial mindset does not allow for waste within the system. Here comes a change initiative email from on high. The senior leaders have identified areas to improve. Leadership modeling may be the next hurdle. Sandrine Provoost hosted a webinar titled *Preventing the 10 Common Mistakes in Leading Your Change and Transformation Projects* (Provoost, 2016). Leadership modeling to her means "leaders not being willing to develop themselves or change their mindsets, behavior, or style to overtly model the changes they are asking." Monkey see, monkey do. The organization will follow its leaders. If the leaders are saying one thing but doing another, no one will follow their change initiative. How can leaders understand the importance of their behaviors during the transition stage? Provoost suggests, "ensure leaders understand the direct link between behavior, creditability, and the success of change." Senior leaders must be aware of how their actions impact change. Assuming they know leads to status quo.

Now that the change ball is rolling, how can we ensure this change remains? Set up a monitoring and reinforcement strategy that includes coaching. Change does not happen overnight. It is a process that survives careers. There must be measures in place to continue the change process. If someone gets stuck, have an outlet for the person to go to. Coaching and consulting are needed for the best to reach their greatest. Look at the stars of any sports team. They have coaches to help get them to that level. No one makes it to the top on his or her own. They may possess the natural talents to be in the conversation, but a coach will tweak the necessary attributes to get the best out of them. The same goes for any profession. Professional development units for certified project management professionals are a way to self-coach. You will learn from other practicing project managers about the ways they do things, what works, what does not work, and how to go about implementing the strategies.

Anything that can be measured can be improved. Set up measurable categories so improvement can be physically recognized. How are you wasting resources? Define the waste. Quantify your goals. Measure progress. Continue the cycle until the goals are met. If you want to cultivate talent, you first must know what that talent looks like. What qualities does a talented developer possess? Is schooling important or is technical skill important? If both are important, which aspects of each one see the most results? Quantifying the change helps employees recognize what is happening. People know what one more line of code per day looks like, what it takes to get there, and can create a plan based on the numbers you give them. Quantify the impact that extra line of code will have on the overall project. How much time/money will that save? Saying we want change and then expecting it to happen is wishful thinking. Maybe your team and organization have the capacity to see what needs to be done, and they do it. Most likely, that is not the case, and some monitoring and coaching will improve the change process dramatically.

Coming up in Chapter 3, I discuss the entrepreneurial mindset. These are ways to take traditional mindsets and flip them to become efficient and effective. You will learn the five actions of the entrepreneurial project manager, the difference between fixed and growth mindsets, and how to make one simple switch that will change your career for the better. All of these techniques can be applied immediately and are results-driven. Chapter 3 is the core of this book.

Chapter 3

Entrepreneurial Mindset

Part of being an entrepreneur means taking risks and mostly failing. Failing should not be a taboo subject to be avoided at all costs. School has taught us that failing means you are bad and cannot handle the subject matter. In life, failing means that you are taking risks and learning. Failing is not important. It is the lessons learned from your failures that determine your future success. Anyone can fail and continue to fail. The goal is to fail, learn, and then succeed. Imagine these setbacks as if you were a rubber band. The farther back a rubber band is pulled, the farther the object will travel forward. If these setbacks are visualized in such a positive light, you will realize successes you once thought impossible. All success stories include failure. It is the classic math teacher sign hanging on the wall: "You miss 100% of the shots you don't take." Without taking, and mostly missing, those shots, you will remain stagnant. Hall-of-Fame baseball players "miss" 70% of the time. That is an incredibly lopsided percentage when applied to business, but the connotation remains the same. Step up to the plate, swing your hardest, and if you miss 70% of the time, that is pretty good. As a project manager, create an environment in which failing is encouraged. During meetings, welcome suggestions that come from out of left field. Do not deter anyone from sharing his or her perspective. A "dumb" suggestion can lead to a discussion in which a breakthrough occurs. An "Ah-ha!" moment can come from someone asking whether this or that is possible.

One way to create that environment is to use improvisation techniques. Think of an improv comedy show you have seen that left your jaw hurting from laughing. Improv comics come up with those laughs at a moment's notice, and not by having someone pooh-pooh their ideas. *Plussing* is an improvisation technique in which you agree to everything using "and" to build on the idea. This technique is based on making your partner look good. Try this at your next team meeting when you are trying to develop ideas or solutions. It will create an environment in which ideas fly around freely, and people who do not normally speak up will volunteer solutions. The number of ideas generated by this practice will be an all-time high. Most of them will probably be useless, but if you find one or two out of 15 or 20 that you can use, this is success.

The goal of this mindset change is to become the "white rabbit" of your organization. I was sitting in on a presentation given at a conference where being the white rabbit was described as a bad thing: They are easily seen by predators and are more likely to end up as breakfast, lunch, or dinner. My immediate reaction was, given those circumstances yet surviving, how awesome is that white rabbit! Think about the mastery of skills that white rabbit has, to survive in nature while sticking out like a sore thumb. All of the white rabbit's attributes must be 10 out of 10. There is no time for the rabbit to relax, as the challenge is around the corner. There is no convenient hiding spot in which to take a break. You have to become that white rabbit—the person everyone in your organization sees and recognizes not only for being different but for delivering results. The rabbit's one goal is to survive. By being alive, these rabbits are doing their jobs superbly. The white rabbit is delivering results by gathering food for survival without becoming food for someone else's survival. Take that mindset into work, into meetings, into negotiations, and so on, to become an entrepreneurial project manager. Stand out from the rest. Get your way. Perform at the highest levels. Be the white rabbit.

Jim Paul, the author of *What I Learned Losing a Million Dollars* (Paul and Moynihan, 2013), wrote: "There are people for places, places for people. You can do some things, and you cannot do other things. Don't get all upset about the things you cannot do. If you cannot do something, pay someone else who can and don't worry about it." Take this mindset to your team. If there is a member of your team who is incredible at certain aspects, give him the work. Do not waste time trying to get people up to speed on tasks that are causing the team to struggle. Schedules need to be followed. Giving someone work she is really good at only makes sense, but it can be overlooked. Not all estimators, programmers, project managers, and so on, have the same competencies. In construction, there is usually an estimator for the dirt work and another for the utility work. These aspects of construction are specialties and have nuances that can only be understood having many years experience. Why would you put someone who does utility estimating on dirt if you have the resources to do it? Some may say to reduce overhead and create a better all-around estimator. That is fine if you can afford the time lost in training and the revenue lost in not being awarded the work. Most companies can afford to have both and should use them to their advantage.

This mindset also applies to subcontractors. The goal of all organizations is to self-perform the work. With that, they are in control of all aspects of the staging and phasing. However, if you do not have the resources to perform certain tasks, hire a contractor that can produce quality results at a reduced cost. Do not look at subcontractors as a liability. They are saving you time and money through their expertise and craftsmanship. If this practice is not the norm, try it. Why not hire a developer to design the website while you create concepts for the brand? Instead of wasting your time trying to learn something, give an experienced contractor the work. I recently did this for a blog I run. I know very little about website design and optimizing traffic. I hired a company to design the website and web search optimize instead of my trying to figure it out on my own. The cost of design was far less than the time it would have taken me to do the same work. Do not be scared off by the initial cost of the subcontractor. Do a cost/benefit analysis. Weigh that with the current workload and make a calculated decision. Your inexperienced 100% is not better than someone else's experienced 100%.

3.1. Stoicism

Don't be ashamed to need help. Like a soldier storming a wall, you have a mission to accomplish. And if you've been wounded and you need a comrade to pull you up? So what?

– Marcus Aurelius, *Meditations*

Ego can build an empire and at the same time destroy it. This quote lends to a later story (Chapter 12) about asking for help off of a "snowy mountain" at a grade school retreat. Why are people so hesitant to ask for help? Does it make us feel inferior? Think about the last time you asked for help and what the outcome was. More often than not, I am going to say the outcome was positive and led to success. Whether it was navigating an obstacle on your path to success or simply asking for directions, you got the answer much quicker by asking the question than you would have by doing the research on your own. Remember back in school when a teacher would say, "There are no dumb questions"? Did that ever prompt you to ask a dumb question? Probably not. Do not allow those childhood experiences to impact your abilities now. A way I have heard people overcome this is to phrase your question by first saying, "This may be a dumb question, but . . ." If that is your way to break the ice, go ahead and use it. In the analogy used in the quote, do you think a soldier is worried about looking weak on the battlefield when asking for help? Of course not. The soldier is worried about survival. Encourage your team to ask as many questions as possible. Ask them questions to ensure that the material is getting through to them.

You do not want to end a meeting thinking your message got across to them and find out in the next meeting that no one did what you expected. Make requirements and deliverables clear and concise. Ask your team members to repeat back to you your expectations for them. Create an environment in which people are asking questions more than making statements. Constant dialogue leads to breakthroughs.

3.2. Taoism

Those who work at their studies increase day after day;
Those who have heard the Tao decrease day after day.
They decrease and decrease till they get to the point where they do nothing.
They do nothing, and yet there's nothing left undone.

– Lao Tzu, *Te-Tao Ching*

How often have you walked past a senior executive's office and it looks as though he or she is doing nothing? Apply this quote next time that happens: The executive may appear to do nothing, yet everything gets done. That's the root of the Taoist approach to project management. Develop your team and cultivate an environment so effective that it appears they do nothing while everything gets done. Does a system get any more efficient? When a basketball team is so good, people think you can just roll the balls out, and the team will win. That is a classic argument when discussing any great coach. These coaches usually have multiple superstars accompanying them in their success, so people will discredit their ability to lead. I argue that it is more difficult to win with a more talented group of players, because managing their egos can be more of a job than trying to win games. If your team has all the top talent, they probably have ideas they are married to leading them to the success they are realizing. When a decision is made that goes against one of their ideas, chaos can ensue. A less experienced team will go along with the flow a little more than an experienced and successful team. We should applaud the manager who appears to do nothing yet his or her team remains successful at the highest levels. There is an environment that is created in which team members take care of the issues before they even land on your desk. That is a result of great leadership, not sitting back and letting whatever happens to happen. The goal for implanting the Taoist belief is to appear to be doing nothing while all the work gets done. Later in this book, I will discuss the Von Manstein matrix, which will further illustrate this point. A busier person does not necessarily mean that more work is getting done. Do not equate busy with productive. I see this mistake made throughout the construction industry. Someone's busy work can make more work for someone else down the line. Think of an operator trying to stay busy while waiting for quad-axle trucks to load. The operator continues to dig outside the project limits because more production is getting done (hauling more cubic yards of dirt). In the end, the operator is creating more work for the bulldozer operator who will be trying to grade the road behind the work. This extension of the right of way causes more aggregates to be hauled back to the project. The sponsor will not want to pay for more material, because the agreement is for the original amount. As a result of someone wanting to stay busy, a chain of events developed later that will end up costing the company thousands of dollars. Do not do work for the sake of work.

3.3. Sales

Grant Cardone, the author of *Sell or Be Sold* (Cardone, 2012), explains sales as "anything having to do with convincing, persuading, negotiating, or just getting your way." This style is the approach entrepreneurial project managers take when discussing sales. It is not trying to gain the advantage by selling someone a dream based on a prayer. Instead of creating buying, you want to create buy-in. Imagine having buy-in from senior management on down to the individual team member. When you bring up an idea or a vision, everyone is on board. Selling the idea is what is going to bring that forward.

First, you have to believe in what you are saying. How many times have you been talking to someone and realized his heart was not into what he was saying? The conversation tends to trail off to nothing of significance. His posture slumps. It is similar to being in a bad relationship with no end in sight. When I was a laborer in construction, I was on the phone with a financial advisor describing my work and what I do for a living. After the explanation, he stated how he wished he could be outside all day and how the job he currently held was not satisfying to him at all. How do you think that made me feel about his ability to handle my money? Do you want someone taking your money who is not interested in what you are buying? What kind of effort do you think he was putting in to find the best investments? The same goes for pitching an idea to your team or senior management. If you have a passion for what you are talking about and your eyes light up when someone broaches the topic, it will show in your presentation. It is obvious when someone truly loves what she is doing. She wants to talk about it every second of every day—looking at all angles of possibility, trying to find ways to improve upon the idea or concept. I am passionate about project management. It shows in my efforts to earn a graduate degree and professional credentials in the field. That is a concrete example of how I bought into the profession. Do the same with your ideas. Do not let them die because someone opposes them. Use that opposition to find new ways of explaining your idea.

Training is important to becoming great at anything. Nobody steps up to the plate and starts hitting home runs. So why is it we expect people to be good at selling without giving them the proper training? Even for project managers, many are thrown into the role and have to learn through trial by fire. Do not make the mistake of thinking selling is easy. Think of the checkout line at the grocery store. Why do you think they put all the candy and magazines right by the checkout? People will buy these things emotionally, without thinking of the psychology of it. The same goes for selling your idea or product. Build confidence, not through luck, but through knowing what you are doing. No one becomes the best by luck. The best train mercilessly in their craft. By reading this book, you are training yourself to become a better project manager. What is the customer trying to say when he or she says the price is too high? It means the price is too high for the value the customer will receive. Maybe the higher-priced item provides more value and is the right choice, as counterintuitive as that sounds. Knowing why comes with training, then experience.

Radical Management^{SM} is a management style developed by Stephen Denning, the author of *The Leader's Guide to Radical Management: Reinventing the Workplace for the 21st Century* (Denning, 2010). The principles of Radical Management, from his website, stevedenning.com, are as follows:

- "A shift in a goal from making money for shareholders to delighting customers through continuous innovation.
- A shift in the role of managers from controlling individuals to enabling self-organizing teams.
- A shift in the way work is coordinated from bureaucracy to dynamic linking.
- A shift in values from preoccupation with efficiency to a broader set of values that will foster continuous innovation.
- A shift in communications from top-down commands to horizontal communications."

While each of those bullet points is important to fostering a new attitude within your organization, I want to focus here on the second bullet point. Bringing teams to the performing stage as quickly as possible should be the main focus for project managers. The sooner they perform, the more effective that team will be. Controlling those individuals will stifle growth and keep them from reaching the performing stage. Because you are a leader does not mean you have to be controlling. There is a difference, and you should be aware of it at all times. The people on your team have talent; otherwise they would not have been hired (for the most part). Allow them the ability and creativity to get work done with the requirements you give them. You may have a certain way of performing a task that has worked for years, but give the person you assigned the task an opportunity to solve the issue on his or her own. This person's way of doing things may help you learn a more effective way moving forward.

Giving your team that space creates an environment in which ideas can flow freely without the fear of getting shot down by a controlling superior. The team will bond over the spread of ideas and learn each other's working styles to complement their own. Most likely, you have adults working for you. Let them be adults. Direction and guidance are required, but *controlling* is a problem. Have you ever taken a day off, and the project has resumed unscathed? If so, you are doing your job. Often, I hear managers saying they can never take a day off because the project will halt without their presence. You are controlling far too many aspects of the project if work will halt in your absence. No one should be that critical, where a day off never comes. Not only are you inhibiting the performance of your team by overcontrolling, you are also causing unnecessary stress to yourself by not letting go. Find a balance. You and your team will appreciate the change.

3.4. Vertical Slices

As project managers, you have many responsibilities and activities on your plate. Whether it is the project itself or your team or the organization, there are always fires to handle. The vertical slices approach takes on each of these issues bit by bit at the same time. Look at Figure 3.1. Imagine each of those layers as an issue. Instead of taking on the problems issue by issue, take a small slice out of each

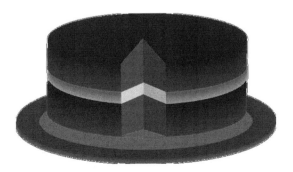

Figure 3.1 Vertical slices.

issue until they are complete. This approach may not be possible with some issues. If an issue arises on the critical path, it takes priority over the others. The vertical slice approach works best for noncritical issues that can be time wasters if they continue to increase.

Take your inbox, for example, physical or electronic. You probably receive many emails per day while paperwork increases. With the vertical slice approach, you will create a dent in both instead of trying to tackle each one until it is complete. Focusing intensely on one task in short bursts is one method to apply vertical slices. Answer emails for 15 or 20 minutes at a time and then take a mental break. Sign invoices or change orders for 20 or 30 minutes at a time and then take a break. After an hour of that, you will have successfully answered emails as well as signed off on paperwork. With this diversion of attention to "new" objectives, it will refocus your energy, and the short bursts will allow for maximum effectiveness in getting your work done. Doing paperwork for hours on end will bore the best project manager. Doing paperwork for 30 minutes can be accomplished quickly and help maintain a sense of sanity over the course of time.

In taking the metaphorical to the actual, what is the most effective way to eat an entire cake? Separating out the layers then eating, or taking a slice of all the layers to consume? Most cakes are eaten by taking a slice of all the layers together until the entire cake has been devoured. The same should go for your tasks. Delegate the remedial tasks so you can focus on the critical tasks. Both will get done in confidence at a much faster rate than if you take on each task and then move to the next. Project managers will fast-track projects but not their issues. The more you can get done by sending one email instead of five, the better off you will be throughout the course of the project. Look for ways to apply this vertical slice approach to your schedule and see the time savings immediately. This approach is a quick win for you as a project manager.

3.5. Six Thinking Hats: Use Them

"Groupthink" has a negative connotation. Team members and managers alike start to think along the same guidelines. Conflict is nonexistent. Edward de Bono, a physician, psychologist, author, inventor, and consultant, decided to use groupthink as an advantage. He developed the concept of the Six Thinking Hats (DeBono, 2000). They are used as a mental technique to redirect your thoughts. The Six Thinking Hats are as follows.

- **Chef/White Hat.** Use this hat first. The chef/white hat (pictured in Figure 3.2) thought process is neutral and objective. It deals with data and facts.

Figure 3.2   Chef/white hat.

- **Construction/Yellow Hat.** The construction/yellow hat (pictured in Figure 3.3) thought process is sunny and positive. It requires optimism, the value of "if," and making it happen. There are no negative thoughts while wearing this hat.

Figure 3.3   Construction/yellow hat.

- **Magician/Black Hat.** The magician/black hat (pictured in Figure 3.4) thought process is the devil's advocate and easiest for most. It asks the question, "Why?" You are exploring the difficulties and dangers.

Figure 3.4   Magician/black hat.

- **Firefighter/Red Hat.** The firefighter/red hat (pictured in Figure 3.5) thought process is the emotional view or gut feeling. It uses intuition or hunches to explore further what is liked and disliked.

Figure 3.5   Firefighter/red hat.

- **Leprechaun/Green Hat.** The leprechaun/green hat (pictured in Figure 3.6) thought process is based on creativity. This stage is to grow the idea. Think BIG. Delve into the possibilities, alternatives, and new ideas. This stage is the best and most fun because it allows the group to run wild with ideas.

Figure 3.6   Leprechaun/green hat.

- **Police Officer/Blue Hat.** The police officer/blue hat (pictured in Figure 3.7) thought process is organizing and thinking. This hat can be considered the project manager role. It is about summary and takeaways.

Figure 3.7   Police officer/blue hat.

Now that the hats are identified, how do you use them? Within your team, pick a hat that the group will wear. Picking a hat will lock the group into a frame of mind. They can only suggest ideas from this mindset. Brainstorm until the ideas run out or as time permits. Use each of the hats accordingly until the project idea is fully determined. Not only does this technique allow for idea generation, it also avoids "analysis paralysis." It keeps each team member thinking about new aspects of the project rather than getting stuck on one or two positives or negatives. The process also allows each team member to get a taste of new roles. As a project manager, you can use this to decide who should fill future roles. You can see who is stronger in different ways of thinking.

As teams perform over many projects, team members become comfortable in their roles. "I am the idea person" or "I like playing devil's advocate." This technique allows for people to branch out and take on different roles without the pressure of being assigned a new role. Devil's advocate is a comfortable role to play because it lets others create ideas while sitting back to critique them. While pointing out flaws and risks of an idea is valuable, getting stuck in that mindset is toxic. Not every idea is bad or harmful. Force your team to see it from an optimistic standpoint. Have them find ways the idea is good rather than bad. This simple trick can add instant value to the conversation.

Josh Waitzkin, the author of *The Art of Learning* (Waitzkin, 2007), was a world-champion chess player. In his time playing chess, he came to realize that quick decisions based on instinct resulted in the best moves. If he took two to ten minutes to decide on a move, he got great results. Poor results occurred when 20 to 30 minutes were allotted to a move. Using his experience and intuition often resulted in his best decisions. Getting into his head and thought process led to poor decisions. Overthinking and not reacting is an issue in project management as well as chess. Your first instinct should get you 80% of the way to a decision. Use analysis to reach the other 20%. The numbers are not scientific, but it proves the point of trusting your initial instinct. Think of the red hat as your North Star while the other hats are the guiding force to ensure you reach your destination.

3.6. Five Actions of an Entrepreneur

In *What I Learned Losing a Million Dollars* (Paul and Moynihan, 2013), Jim Paul discusses the idea of created or inherent risk. He relates it to the potential loss of money in the markets because his profession was a trader on Wall Street. I would like to take his activities and apply them to project management. Any time project managers take on a new project, there is risk involved. Inherent risk is the probability of loss arising out of circumstances or existing in an environment, in the absence of any action to control or modify the circumstances. Inherent risk is always present, and project managers put their best foot forward to control it. If control is not enough, there should be steps in place to take action to limit the impact. Created risk is an action to bring about the risk that is not a natural by-product of the activity. You are taking on additional risk by performing a task. What determines inherent versus created risk is the activity you are performing. Following is a brief discussion of categories of activities that investors, as well as project managers, partake in.

3.6.1. Investing

Projects are normally investments. An organization parts with its resources expecting a return on said resources. The return is periodic payments in the form of interest, dividends, or profits. Good investments result in the highest returns. Poor investments do not keep an organization in business very long. Project managers invest time with the expectation of production. After a day's work, the project should be further ahead than when the day started. The return on your time investment is progress. Project managers also receive monetary returns in the form of a paycheck. Your investment of time should reflect the return you receive. The same goes for the project. The resources invested should be worth the return. Investing takes on a long-term view. Returns may not be immediate.

3.6.2. Trading

In the financial world, a trader makes a market, buys at the bid price, and tries to sell at the offer price. It is similar to flipping a house. The goal is to create a market for this run-down shack by fixing it up and reselling it for profit. These projects come with higher risk because of the longer list of unknowns. The house may appear decent on the outside, but tearing down the first wall may reveal some expensive surprises. There are countless television shows based on flipping houses. Numerous times an older home leads to rewiring electrical or rerouting plumbing. Both issues cannot be seen from an outside perspective. It is an inherent risk you are taking on as a project manager. You are creating risk by tearing down the walls to create a better market for the home.

3.6.3. Speculating

Speculating is parting with capital with the expectation of capital appreciation. A purchase is made based on resale, not use or income. Speculating requires an intellectual examination. This technique tests your knowledge of the market and its future potential. You are purchasing an item, tangible or intangible, with the hopes of its value increasing over time, to resell for profit. Speculating is different than investing because it does not have periodic returns such as dividends or interest. Speculative projects offer uncertain returns. The project team puts in the research, develops a business case, and the project gets approved. However, there could be unforeseen market shifts that derail the entire project. Energy companies are a great example of speculation at work. Exploration projects tie up a majority of their resources in hopes of hitting a new vein of energy. There are many failures before success. Speculative projects have high risk but can deliver high rewards.

3.6.4. Betting

Betting is a game of right and wrong. Two parties agree on a reward, usually money. The wrong party forfeits the reward to the right party. Betting on projects is never a good idea because it puts egos on the line. There is always a clear winner of a bet. A football team loses. You thought they would win. You lose and forfeit the wager. Betting within a project takes place more among individuals than on an organizational level. In construction, production is important. The entire project is based on how much you can get done in a day. The estimate was based on production. If you produce less, the organization loses money. When a crew was outperforming the estimate, a friendly way to challenge them was to bet a steak dinner. If they could attain a doubled goal (say 10,000 tons crushed rather than the bid quantity of 5,000 tons), the foreman would get a steak dinner from the boss. A $100 steak dinner was worth thousands to the organization yet motivated crews to attain the goal.

3.6.5. Gambling

Gambling and betting are closely related. The difference comes in the form of entertainment and chance. Gambling is not often about being right or wrong. The odds can be stacked in your favor, yet you still lose. Your action was right, but the outcome was not a win. Gambling takes place on projects. Building demolitions are a great example. You thought there would be more copper in the demolition. In your proposal, you gambled that money to make up for work packages you had to come in at cost. Not only was that work package bid at cost, you are also performing for a loss. The odds may have been in your favor. Based on your experience, copper always plays a factor, but in this instance, its impact is not as great. Your gamble would not pay off. The opposite is also true: The odds may be stacked against you, yet you strike gold.

Knowing these activities will impact your management style. Are you creating risk? How much inherent risk are you taking on? Does the benefit outweigh the cost? These questions are answered differently based on the action taken. All activities come with risk. There is no way to avoid it. Inherent risk can and should be controlled. These risks should be known. Created risk is something a project manager can limit. Do not make things worse by gambling on investment. Realize which of the five actions (investing, trading, speculating, betting, and gambling) you are taking on and act accordingly.

3.7. New School: Kicker and High Jumper

In Chapter 2, I discussed traditional ways to kick field goals and high jump. This chapter deals with entrepreneurial approaches to those traditional systems. Pete Gogolak was the first Hungarian professional football player and brought his soccer roots to the field. Instead of the straight-on approach using the toes

Figure 3.8   Pete Gogolak (Gogolak, 2016).

for impact, he approached the football at more of an angle, using the instep of his foot to kick the ball. This approach may seem obvious to any viewers of today's game of football, but in his time, it was revolutionary. No kickers had been doing this before Gogolak introduced the league to soccer-style kicking as shown in Figure 3.8.

As you can see, it is a slight variation on what George Blanda was using successfully. This new approach to kicking a football is now used throughout the NFL. No current kicker uses the straight-on approach. The best practice of yesteryear has been replaced by today's best practice.

Dick Fosbury introduced the world of high jumping to the "Fosbury flop," as shown in Figure 3.9. Using the newer raised, softer landing areas, his technique of leading with his head and shoulders while jumping backward over the bar earned him the gold medal at the 1968 Olympic Games.

Not only was he taking advantage of his technique, floppers around the world began dominating competitions. The Fosbury flop is the technique we see today in all high jump competitions. It would be alien to see any other technique used in competition. Again, it was a small variation of the techniques already being used. This change in best practice began setting new records and smashing the competition.

Revolutionary ideas are not a complete 180-degree difference from previous best practices. The subtleties can make all the difference. Gogolak and Fosbury

Figure 3.9   The Fosbury flop technique (Fosbury, 2016).

both had a challenge in front of them. Both techniques take a slightly different angle to their approach. Instead of using the same approach, they tried a slightly different method that became the new standard. This standard was set by experimentation and by trial and error. Take a look at the systems you have in place. Are they the best? If not, why? If you think they are the best, how can they be improved? Play around with different ideas. Remember, failing is learning.

3.8. Fixed and Growth Mindsets

Dr. Carol Dweck, the author of *Mindset: The New Psychology of Success* (Dweck, 2006), developed two mindsets for learning: fixed and growth. A fixed-mindset individual will choose activities to confirm his or her ability. A growth-mindset individual will choose activities to expand his or her ability. Dweck noticed that how students perceive their ability to learn influences their achievement. When students learned they could "grow their brains," they outperformed the other students. Students' thought process about the tasks influenced their outcomes. I want to extend this ability to our teams and organizations. There can be a culture of this is how we do things, and that is why we do them. The scheduler does scheduling because schedulers schedule. Of course, that is true, but your team is not growing when that is all this person does. You are giving this individual tasks that confine his or her ability. That may be the person's job, but mix in some tasks to expand the person's ability. Show the scheduler how an estimator performs. Knowing more roles can help the team collaborate more effectively.

If the estimator knows how the schedule is compiled, that may influence how the estimate is put together. Full days might be better for the scheduler than by the hour. Now, the scheduler's job becomes easier, and it allows this person to pick up more demand. Neither party needs to become a master in the new skill, but gaining that small bit of knowledge into other worlds can help the team be more effective and efficient.

How can we implement a growth mindset into our team? Below I have included actions and approaches that trigger growth and fixed mindset.

- Fixed Mindset
  ○ Challenges are to be avoided. Instead of seeing them as opportunities, a fixed mindset would rather avoid them or give up quickly, to end the growing pains.
  ○ Criticism is seen as an attack rather than a building tool. Criticism remains personal, never professional.
- Growth Mindset
  ○ Take on challenges. If the challenges become difficult, instead of giving up, a growth mindset finds ways to power through the setbacks.
  ○ Criticism is seen as a way to develop skills and become better.
  ○ The success of others is inspirational, not a threat. Growth mindset wants everyone to be successful and raise the bar overall rather than just themselves.

Again, these are about changing mindsets, and words do matter. Brushing something off allows that mindset to stay active. These mindsets are not about letting people off the hook because they are not good at something. It is identifying a weakness and working at it. Technology is a huge part of our day-to-day work, yet I see many project managers refusing to use it because the learning curve is steep. They completely abandon technology, hire a person who understands it, and both continue to perform fixed-mindset activities.

Dweck points out a common misconception of the growth mindset. A growth mindset is not just about effort. She relates the difference to students being rewarded for effort but not learning. It is like carving pumpkins with my father at a young age: I would saw and saw and saw, eventually leading him to take over because he could do it better. I did not learn anything by having him take over and do it for me. I put in the maximum effort until my arms were sore. All I learned is that it is easier for my father to carve than me. I would get relegated to removing the seeds and gooey middle while he would carve.

Participation trophies come to mind as well. There is something to be learned by losing, even getting blown out completely. Humbleness is restored. A luster is returned to winning. If you get the same trophy for coming in last as you do for coming in first, you are learning that to show up is important but how you perform does not matter. How does this apply to your team? Are people getting recognized for showing up instead of for performing well? Is effort being rewarded, rather than learning? The goal is to develop your team through activities that challenge them. Push their mental limits and reward based on the process, not just effort.

In Chapter 4, I discuss the Von Manstein matrix, which shows you where effort belongs in the organizational chart. The answer may surprise you. I also take an in-depth look at philosophy and sales to improve your organization's strategy moving forward. I discuss common mistakes and how to avoid them within your organization. Are you a gym teacher or a coach? This question will harken back to the days of high school football. I also discuss followership, a word rarely used yet so important to the success of projects and organizations. Not everyone can be a leader. Even if you are considered a leader, you must sometimes take a step back and follow. There are also techniques for executing your strategy and keeping the relevance of your strategy at the forefront of your team's mind. Chapter 4 takes a holistic approach to your organization and its strategy.

Chapter 4

Organizational Strategy

This chapter presents a story about how not to handle a resolution to an issue. For example, assume there is an issue out in the field, and it is brought to the attention of the project manager. The project manager lets the senior manager know about the issue and tries to resolve the issue as a duo. In the end, the boss's resolution works, while the project manager's resolution ends up not working. In a meeting later in the week, the issue is brought up as a learning tool for the group. Instead of explaining the issue and how to resolve it, the senior manager decides this is a time to stroke the ego. The boss asks the project manager his thoughts about the resolution, knowing full well the project manager's solution did not work and the problem could not be resolved without the help of himself, the senior manager.

Do not call out your team members in public, especially to show that you were right about how to resolve the issue. If you want to bring up the issue, keep the information general and nonspecific so the group can learn. Present the resolution by fact. There is no need to make yourself the hero of the story. The goal was to resolve the issue, and once that was taken care of, let that be the end of the story. Keep lessons learned to the facts, without personalizing or internalizing the success. Your team does not care who resolved the issue; they care that it has been resolved.

4.1. The von Manstein Matrix

Erich von Manstein was a German commander during World War II, eventually attaining the rank of field marshal. He held former German general,

Commander-in-Chief of the Reichswehr Kurt von Hammerstein-Equord, in high regard, stating that Hammerstein-Equord was "probably one of the cleverest people I ever met" (Street, 2014). Both men have been credited with the development of the Von Manstein matrix used to select officers. While this a German military grouping, it applies to project management and team development as well. The matrix places team members into four categories. Those four categories are smart, dumb, hard-working, and lazy. I use the extremes of intelligence and initiative for effect; I do not mean to imply that the person placed in the "dumb" category is really dumb. The von Manstein matrix is shown in Figure 4.1.

		Initiative	
		Hard Working	**Lazy**
Intelligence	**Smart**	Middle Management	Promote Them!
	Dumb	Fire Them!	Leave Them Alone

Figure 4.1   The von Manstein matrix.

According to the von Manstein matrix (Street, 2014), the dumb–hard working need to be dealt with immediately. They cause problems where problems should not exist. They are problem solvers who create more problems than they solve. Dumb–hard-working people tend not to see the big picture. Instead, they can create a domino effect of issues.

The common viewpoint might be to fire the lazy–dumb crowd. Who would want someone lazy and dumb? The key with these individuals is that they keep to themselves and do their jobs. They are the "no harm, no foul" group. Dumb–lazy may make up most of your team members. Routine duties are perfect for this group.

Smart–hard working may jump out as people to hire all the way to the top. Instead, their compliance allows them to deny responsibility. They are letter-of-the-law type team members who cannot see opportunities because "that is not how we do things around here." Smart–hard-working individuals are safe, reliable cogs of the machine. Also, the hard-working mentality does not allow for reflection. They have their heads down, plowing through work.

Finally, smart–lazy is the sweet spot for your upper management. Von Hammerstein-Equord said that smart and lazy is qualified for the highest leadership duties, "because he possesses the intellectual clarity and the composure necessary for difficult decisions" (Street, 2014). They are problem finders, not problem solvers (discussed more in Chapter 5). Think of lazy as "work smarter, not harder." These people would rather find a better way than continue the status quo. I have experienced so many senior managers who are smart–hard working. They have the technical background to succeed on the operations side but lack the clarity for long-term decision making. They are also so busy working hard that they do not have time to look for resources to better their weaknesses. They get stuck and cannot find time to get out of the rut.

Find out how your organization does business with its clients and stakeholders. The construction industry is based on unit pricing for quotes. The lowest bid package is awarded the procurement. If that is not your style of doing business, construction is not the industry for you. The relationship develops in the price, not the person. Even if you have worked with a company for years, they are looking for the best price on the market. If your price comes in higher than a competitor's, you will probably lose the business. Some companies do "Invitation for Bid" exclusively to build those relationships, but most send out a mass invitation and choose the lowest bidder. The bottom line becomes a priority. Bidding is a cut-throat process. Bidding near cost and then finding ways of making money throughout the project has become the norm. Margins are slim.

Throughout an organization, there should be some level of uncertainty. If people become too comfortable in their roles, performance decreases. When I worked for family businesses, the comfort level reached levels unmatched anywhere else. Once people find their niche, they tend to settle into that role and do not aggressively pursue other avenues. Set a tone of unease within your team. Have them constantly strive for greater heights. Involving family complicates that dynamic. No one wants to fire a relative because of poor performance. That does not mean everyone is on edge at all times and fun is seen as a deterrent. Use that unease to motivate your team. Keep things interesting by mixing up roles in meetings. Have people perform different tasks on different projects to see what they take a liking to and create a better all-around performer for your organization. The more experience a team member gets, the better performer the team member becomes.

A webinar titled "Organizational Project Management Maturity Is Project Management Quality" (Quiring, 2014) stated: "Quality improves when the organization as a whole is educated and understand their role and matures." It is vital that each team member understands his or her role and abides by the rules for that role. How many times have you encountered someone stepping out of his or her role and causing confusion? It is an inevitability that occurs on project teams. Someone wants to be the leader but does not have the title for that role. There can only be so many superstars on a team. Role players are the glue that holds the team together. They do the dirty work. Tristan Thompson is a role player for the Cleveland Cavaliers. His job is to defend post players and grab rebounds. Anything outside of that (scoring, assisting, or stealing) is icing on the cake. If he wanted to be the one to score all the points, it would be a detriment to the team. The Cavaliers have players on the team to score, most notably LeBron James. Because of Tristan Thompson's ability to fit his role, he was paid $82 million for five years. The significance of the contract is that he gets paid like a superstar for filling a supporting role. That is how important it is to build a team of not only talented individuals but also people who recognize their skill sets and play to their strengths. It does not do anyone any good if a team member continues to struggle with a task that someone else does exceedingly well. No one gets better. The team grows frustrated, and production suffers.

That ability comes from education and maturity. At one point in time, we all think we are going to be the best at something. It takes maturity to recognize what you are good at and, most important, what you do poorly. That maturity comes from education. Learning along the way. Taking risks. Falling and getting back up. Without those experiences, we will not learn the distinction and continue to *think* we are good instead of *knowing* we are good.

4.2. Stoicism

> *Love the discipline you know, and let it support you.*
>
> – Marcus Aurelius, *Meditations*

Take a look at your organization. Do people love what they do? Or do they go through the motions, doing just enough to keep their jobs? If people love what they do, it will show in their work, attitudes, and performance. Continued learning is a dead giveaway that people love what they are doing. Once people are taking steps outside of work to gain more knowledge and understanding of a discipline, it is a true sign of passion for the discipline. Not only acquiring degrees or accreditations, but also reading the latest news and talking with their peers about upcoming developments. People who love what they do cannot talk enough about what they do. Think about people you know who have started a new diet or training regimen. All they ever seem to want to talk about is how much better they feel being on this new diet or how effective their new trainer is, because the trainer is showing them things they never knew before. People can talk about their passions for hours on end without growing tired of the conversation. How many times have you gone to lunch with your team, and all they talk about is work? That is a sign of how involved they are in the project, and the interest it is generating. Even on their own time, they continue to talk about what is going on or what they plan to do next. Continue to foster those conversations with open dialogue, and add your thoughts to the conversation to fuel further discussion. The only times I do not want to talk about work is when the project being discussed does not interest me.

When you love what you are doing, that discipline will support you financially and professionally. People want others around who share the same passions. Let that love of the game support you in good times and bad. The worst day at a place you love is better than the best day at a place you hate. The discipline we have chosen to love is project management. I chose project management because of the broad spectrum it covers. There are endless possibilities when it comes to projects and how to manage them. I started in the construction industry and have since dabbled in the online space as well as working on IT projects. The widespread project management offerings continue to present new challenges and creative ways of solving those issues. The discipline has supported me outside of work as well. When real-life challenges come up, project management techniques can often be used to overcome those issues as well.

Communications are the main link to solving most issues. Running your life like a project can help to improve situations you never thought possible. Most people try not to take their work home with them. Think of the electrician who comes home: What is the last thing he wants to do after work? Fix any electrical issues around the house, since that is what he does all day. Let your passion for project management shine both at work and outside as well. Planning birthday parties is a classic example of an important project for your most important stakeholders, your family. Throwing the best birthday party for a middle schooler can be a life-changing event.

4.3. Taoism

The more knowledge and skill people have, the more novel things will appear.

– Lao Tzu, *Te-Tao Ching*

Pay close attention to how people talk about projects within your organization. Do they make the projects seem mundane and everyday? If so, your employees have knowledge and skills that are superior to the projects they work on, and they can be challenged to reach greater heights. When describing projects, do you come off as sounding bored with the topic? Challenge yourself to improving the way you do things. In construction, there are only so many ways to build a road or crush aggregate. At some points, describing what you do to another person can be a novelty. We take our industries for granted because it is what we do. Other people do not understand what we do, and those people can even be stakeholders. On a construction project, stakeholders include residents of neighborhoods where the street is getting replaced. They do not understand the methodology for replacing a road. All they see are orange barrels, heavy equipment, roads and yards dug up, and the like. They want the equipment and machinery gone and out of the way. They do not care how it gets done—just the sooner, the better. If you go on describing this construction as a novel idea to them, they may get upset because you are causing them an issue by tearing up their driveway. Be cognizant of the fact that you are important to these people, and assure them you know what you are doing. Amplify your knowledge to them. Use the above quote to let them know you have the knowledge and skills necessary to complete the work.

Be cautious of the individual who is describing something in great detail to a group of professionals in the industry. That person is trying to convince the group of his own depth of knowledge. A group of project managers probably does not need a detailed talk about what a project charter is or how to go about designing a work breakdown structure. There is a common understanding among the group of how to do the foundational aspects of project management.

4.4. Sales

If your organization has a sales department, it is a great resource to ask questions and figure out how to sell. Pick out the top salesperson. Take him or her out to lunch. Pick this person's brain. You will soon realize all the intricacies that go into making the sale. When I was describing project management to someone over dinner, he made a comment: "I did not realize all that was involved." If people have not had the first-hand experience, they will create a story in their heads describing what they think happens. Experience takes that "think" and makes it "know." That is what the lunch with a salesperson will do if you are unfamiliar with sales. Things the salesperson takes for granted are things you did not knew. Ever hear the phrase, "He's forgotten more than you will ever know"? It is a sign of a true master in the industry. Also, people want to share their story. Asking questions is the easiest way to get the answers. Sounds obvious, but it is often not remembered.

Some organizations even have a training program for this. One company I worked for called their system a university. It provided training on the company intranet, GPS systems, software, and so on. Take advantage of these opportunities. Allowing your team members to get paid training only makes your team better. A speaker was coming to our office to talk about better business etiquette. In the posting, it was shown as an unpaid learning opportunity. Right away, people were discouraged to attend because they would not be paid for their time. Is $20 per person (assuming their hourly rate) not worth the betterment of your team? I would make a point to be sure they get paid while learning. In trying to save a little money, you are being a detriment to your team by not encouraging them to learn and become better at their jobs. There are plenty of organizations willing to risk hundreds of thousands of dollars on an insight rather than spend that money on their employees. The return on investment by investing in your team is ten times the return on any project you assign them. Think of the impact a book or a speech or a talk had on you, and the effect it continues to provide. These opportunities should not be overlooked. Having a strong team is not a threat to your leadership. You will not be replaced because someone on your team is stronger than she was before you sent her to the talk. It will enhance your leadership ability. Dealing with smarter team members is easier than the opposite. An organization that labels an opportunity as unpaid eliminates most of the interested parties. I know I was immediately turned off at that point, even as someone who is constantly trying to learn new things. I cannot imagine how people who are not interested in learning feel about those opportunities. The postings are probably deleted without ever opening them. Encourage continuous learning in your organization. Be the guiding light for material and where to look for information. It will not only make you a better leader, your team will be stronger.

If you believe you are using best practices, you are not getting better. The organization may have best practices, but they can always be improved. The statement, "We use best practices," shows an ending to the maturation process in how your organization does things. Because they are the best, they do not need improvement. Best practices give a false sense of doing all that you can to perform the work ahead. Constantly improve the way your organization and team perform work. "Why?" can be the hardest question to answer. Do not assume that best practices are being used because someone labels them as such. Going through the motions impacts not only your organization but also your team's performance. They will get stuck in a lull. Having your team question best practices is a good way for them to think about their efforts critically on a day-to-day basis. Later in this book, you will learn about the S.C.A.M.P.E.R. technique for a systematic way to question everything.

4.5. Common Organizational Mistakes (and How to Avoid Them)

The following sections are based on a presentation by Dana Brownlee titled "4 Common Rookie Project Manager Mistakes and How to Avoid Them" (Brownlee 2015):

4.5.1. Ignoring the Slacker

Have you ever been on a team where one or more person(s) were not holding their weight? Of course. School always comes to mind. There would be a group of four individuals meeting at the library the night before a project is due, and one of the students would be on the phone or scouring the Internet while the rest of the group did the work. In the end, everyone got the same grade, because everyone's name was on the presentation. In a scholastic setting, you are stuck with that individual. Not the case in the real world. Do not allow this person to get away with it, for whatever reason. The organization must build a culture of accountability. More important, the team must create that environment.

The three roles each team member can play are *bull, thoroughbred,* and *lamb.* A bull is task-oriented, one who is focused on the goals at hand and is willing to get them done at all costs. A lamb is relationship-oriented, one who is focused on team morale and knowing people on a personal level. A thoroughbred combines the best of both worlds, one who is focused on the tasks and goals of the team while keeping morale high and building work relationships. You, as the project manager, need to be the thoroughbred. Your responsibility is to keep the team on task while creating an uplifting environment in which to work. The bull side of you will allow you to stay strict with the team and moving forward. The lamb side of you will help your approachability should a conflict arise, or if a team member has an idea he or she wants to run by you. A 50/50 balance is unlikely, but get as close to the center of the spectrum as possible.

Questions to ask a potential slacker are as follows:

- What is your understanding of the task?
- What does the deliverable look like?
- What are your first three steps to accomplishing your task?

These questions will ensure an understanding from your team member and clear up any confusion as to what is expected of him or her. Asking in a team meeting is most beneficial. Not only does the slacker have the pressure of the team, but also it encourages the other team members to ask themselves the same questions without the spotlight. Create a thought process for all of the team members. They will question themselves along the way as to their expectations. Give them opportunities to ask for clarifications. Laying out some tasks, assigning them, and then exiting the meeting only adds to the confusion. The three questions, if answered correctly, show a deep understanding of what your expectations are. Ignoring the slacker only compounds the problem. It will not take care of itself.

4.5.2. Skipping the Initiating Phase

The initiating phase's main output is the project charter. Without the project charter, there is no project, yet organizational leaders want to get right into planning the work. How are we going to get this done? What resources are we going to use? What does the schedule look like? None of those questions can be answered unless you know what you are talking about. That may seem to be an obvious statement, but it needs to be emphasized. Fully vet the project before starting to plan. An organization that is willing to plan a project knowing only high-level details is taking on unnecessary risk. Take the time to gain a full understanding of the project and potential stakeholders. Even if you have worked with the stakeholders in the past, make sure your contact information is up to date and check their recent performance, because changes occur over time. The stakeholders may have the same name, but do not assume they bring the same skills to the table.

4.5.3. Skipping Risk Analysis

Risk analysis is not a luxury but a necessity. You need to perform a proper analysis. Similar to properly vetting the project before you start to plan, you must fully understand as many possible risks of the project. Develop ways to handle them. Positive risk responses are share, enhance, and exploit. Negative risk responses are avoid, transfer, and mitigate. Not only do you have to identify the risks, you also have to rank the risks and then provide responses for each of them. Knowing what to look for is a start. Knowing their potential effects on the project is better. Knowing how to respond to those risks and their effects are best. The math to decide their effects on the projects is simple: the probability of the risk occurring multiplied by the impact the risk would have the project should it occur. Take the qualitative risk and quantify it so the team understands why this step is important. Saying that a risk is really, really bad does not come across as strong as saying that this will set us back one month and cost $20,000. Perform an exhaustive risk analysis and put together a comprehensive list showing what the risk is, how great an impact it will have the project if it occurs, and what the response will be if the event happens.

4.5.4. "Sugar-Coating" Status

Project managers experience stress in managing a project within schedule, scope, and budget. When your boss comes up to you and asks nonchalantly,

"So how's the project going?" it can be easy to sugar-coat by saying "Good!" and then walk away or come up with a list of reasons why things could be going better. Do not sugar-coat the status of your project. If no one knows the true status of the project, adjustments cannot be made that might correct the issues. Maybe you need one more person on your team, and someone has an extra team member he can lend you for a few days, but if you do not make anyone aware of your problem, no one can offer a potential solution. Another way I have heard it described is "ugly baby syndrome." Parents always think they have the cutest baby. No one will tell people their baby is ugly, because that is impolite and rude. The same goes for projects. Because you are the project manager and have a close relationship to the project, it may seem better than it is. Give yourself the outside perspective. Take a look at the success criteria you established in the project charter. Are you meeting your goals? Prioritize the triple constraint. Objectively analyze your project and relay the information to your team, senior managers, and even stakeholders. The cover-up is always worse than the crime.

4.6. Gym Teacher vs. Coach

While listening and reading about the topic project management, I have come across many people who have been thrown into the role of project manager. They have worked for the company for some years, excelled at their position, and their boss wants to promote them to manage a team. This individual has no formal project management training and may have only reluctantly accepted the position. On the flip side, there are individuals who have studied project management, earned the credentials, and accepted a role as project manager. They have a desire to be a project manager and continue to learn about the subject matter even after acquiring the position. The comparison between the two is similar to hiring a gym teacher to coach a sports team instead of someone who is or wants to be a coach.

Colin Cowherd is a popular sports radio host. His sources were lamenting the fact that college football has turned into a game of recruiting. Whoever lands the best talent should win the most games. Assistant coaches and coordinators are being promoted to head coach because of their ability to recruit the top talent rather than their ability to coach a football team. The quality of coaching in college football has suffered. Relate this to my scenario above. The person getting promoted to project manager because he or she is a really good developer will lack the expertise it takes to be a project manager. However, the person with the training, continued education, and desire for the project manager role will be more qualified and should provide a better experience. Organizations want to hire coaches, not gym teachers.

There are ways to become a coach if you are a gym teacher:

- Read books about the subject—if you are reading this book, you are already ahead of most gym teachers.
- Watch webinars and other educational material—not only on project management, but also on leadership, career development, your industry, and so on. Above, you are talking about the gym teacher—is this what you want?
- Observe—look around at other project managers in your organization. How do they perform their work? What tools and techniques do they use for success? What words do they speak?
- Write and document—see what works and what does not. Once you have a feel for your team and your organization, you will notice tendencies in the way people operate. Use those to your advantage. See what makes individuals tick and how they respond. Write those things down and look back when the time comes.

Gym or health class teachers were always coaches at my high school. They have some relation to sports but not a full interest. You want to be at full capacity as a project manager. All in. Being a gym teacher is acceptable if that is your job, but being promoted to coach has an entirely new meaning.

4.7. Project Management Office Maturity

Some organizations have a project management office (PMO) to handle all project management–related activities. PMOs have templates for documentation, including project charters, risk registers, work breakdown structures, and so on. They strive to standardize how a project is executed across all platforms and maintain those standards throughout the organization. The maturity of the PMO is important. It helps to align project outcomes to strategic priorities. The more mature, the more standardized the operating procedures are, making it easier for project managers to do their job. Three types of PMO maturity are process, people, and technology.

Process is defined as a series of actions or steps taken to achieve a particular end. In this case, the end refers to a deliverable established by the project charter. If the organization's PMO is highly mature, there will be a standard in place that describes how best to deliver a successful product. A procedure is described through documentation, and ways to handle conflict, should it arise, also is documented, to see what worked best and why. For example, assume you have a change order. The PMO should have a process for how to handle it. Who do you tell? Who needs to sign off on it? When can you perform the work? How will it be paid for? All of these terms and conditions should have a process and a template. If your organization does not have a PMO or it is immature, there may not be a standard for change orders. A napkin and a crayon may be sufficient.

People refer not only to people within the PMO but also within the larger organization. Do you have coaches or gym teachers? Is your staff qualified for the positions they hold? Are there processes and standards to follow, or do the project managers go about their business however they want? While the paperwork and documentation may get cumbersome at times, it exists for a reason. A mature PMO will not only hand out the documentation to be filled out, it also will follow up to make sure it is being filled out properly. Garbage in, garbage out. The standards and templates need to be filled out correctly; otherwise, the information received will be useless. An immature PMO will take the paperwork and file it in a drawer. A mature PMO will look at the paperwork, make sure the details are sufficient, and file it away in the proper project folder where it will be easy to locate later. The people of the PMO are the most important part. They need to buy into the systems in place for it to work.

Finally, *technology* plays a role in the PMO's maturity. What systems are in use? An electronic form with drop-down menus is easier and is quicker to fill out than doing it with pen and paper. Are the templates being placed in a physical folder within a drawer system, or are they being cataloged electronically for easier search capabilities? I worked for a company that had a basement full of boxes containing paperwork 20 years old. Someone would write the dates on the front of the box and put it high on the shelf for a referral. When a person needed something, the person would go to the basement, look at each box until the timeframe was located, open the box to a slew of scattered paperwork, and search each document hoping to find what he or she wanted. Does this sound like 2017 or 1957? Technology needs to keep up with the times. If you do not have the technology for electronic forms, at least scan them into an electronic folder for keyword search capability. The PMO's maturity will improve the way project management is done for your organization. If a PMO does not exist, take these techniques to your team. Make sure they are up to date on their skills, and keep documentation for your team. If they need a template or a standard procedure explained, be a good resource for them.

4.8. Followership

Leadership is often emphasized within an organization. How do we become better leaders? On the other hand, followership is often overlooked. A webinar titled "Leadership is Half the Story: A Fresh Look at Followership, Leadership, and Collaboration" (Hurwitz and Hurwitz, 2015) by Dr. Marc Hurwitz and

Samantha Hurwitz delves deep into the topic of followership. They start with an experiment performed by John Dyer and his team at Leeds University (Dyer, Croft, Morrell, and Krause, 2009). He took groups of four guppies that were classified as all leaders, all followers, and mixed groups. When the groups were foraging, the all-follower groups performed the worst of all. The all-leader group did slightly better than the all-follower groups. However, the mixed group vastly outperformed both of the other groups, showing that both types of behavior are needed all of the time.

While leadership is important, without followers, there is no leader. Being a follower has a negative connotation as someone who cannot think on his own and needs to be told what to do. With the rise of social media, the follower has taken on a different message. The follower is seen almost as a friend rather than someone who follows. If you follow someone on social media, you become part of their lives. You can interact with them. Socialize with them. Without this interaction, the leader is talking to no one. There has to be a leader–follower dynamic for both parties to get anything out of the experience.

Even if you are a project manager leading a team, you have to follow. More than likely you have a boss who gives you instructions as to the next move or moves. Your followership of your boss translates to the leadership you provide your team. If you follow your boss's instructions poorly, you will lead your team astray. You will perform poorly as a result of your followership, not your leadership.

Followership gets rewarded more than leadership. Think about it. Your boss gives you a raise based on how well you follow his or her instructions. If you decide to go a separate route than your boss (poor followership), conflict will ensue no matter the outcome. Same with your team. Who do you go to at crunch time? Someone who listens to you (great followership), or someone who hears what you are saying but does not take action (poor followership)? Of course, you go to the team member who exudes the best followership.

There must be a commitment from leadership and engagement from followership to strengthen the leadership–followership dynamic. As a leader, you must be committed. This commitment includes what you do personally and your actions toward a project or idea. You must also show your engagement. This engagement is what others see. If you are leaving early but still doing work at home or on the road, some people will question your commitment because they do not see it for themselves. The same goes for followership. You must be engaged. Show up on time. Take notes in meetings. Help others. You have to show your engagement as well. Doing all of those things behind closed doors is not showing your engagement to leadership. They have to see it to believe.

Also, think of followership as a form of leadership. As an assistant project manager, I would be given tasks without guidance. The goal was to get the task done and report back to the project manager. With good followership comes great leadership. Ralph Waldo Emerson wrote, "Do not go where the path may lead, go instead where there is no path and leave a trail" (Emerson, 2016). When you are following yet have the ability to take the task wherever you think is best, use this control to your advantage. You are controlling the pathway and how something gets done. Your boss may give you tips on how to go about it (established path), or you can take on the challenge in your way (new path). The creation of new paths shows leadership in followership.

4.9. Execute Your Strategy

Your organization has a strategy, and now it is time to implement. Whether the strategy is old or new, you have to implement it with new hires or current staff. A webinar titled "Uncover 3 Hidden Change Levers to Successfully Execute Your Strategy" (De Flander, 2016) by Jeroen De Flander will help us do this effectively. He talks about using the "3-H approach": head, heart, hands. For your strategy to work, people need to be aware (head), need to care (heart), and need to do (hands). Most organizations focus on the head. They make people aware of the strategy and expect management to implement. This approach to implementation is wrong. They need to start focusing on the heart. If you get people to care about the strategy, they will naturally be aware and start to do.

So how do we get people to care? Stories for the heart are like food for the stomach. The heart loves a good story. I remember listening to the radio with my mother for Children's Hospital Radiothon to donate money. Every story they would play about parents going through their struggles would make my mother want to donate more and more. These stories played to her heart, which helped to implement their strategy of saving lives. This approach can work for you too. Telling people to be safe in the field is not nearly as effective as having someone come in who has lost a loved one to an accident at work. Seeing someone in emotional pain from their loss will get the message across better than showing them statistics on workplace injuries or a presentation they have seen the last three years. Stories build that emotion inside of them and make them think.

De Flander mentions the "E" experiment. He tells the audience to draw an "E" on their forehead. Which direction did you draw the "E?" A self-oriented "E" (meaning you drew the E backward for the audience) is three times more likely for high-power participants. This orientation means that if you are in a position of power in your organization, you are drawing a backward "E" for your audience, which translates into poor communication. How are you supposed to get your message across if your message is directed at yourself? If the story you are telling is for yourself, your audience will lose interest. They need a message directed to them. Use this "E" experiment as a self-test for your mind-set when delivering a message.

4.10. Relevance and Meaning

Relevance and meaning refer to projects leading change in your organization. Organizations will get their marketing teams to pitch the next great idea to run rampant through your organization. Whether it is leadership training, a policy change, or reorganizing the organizational chart, change is imminent. Sandrine Provoost (Provoost, 2016) gives us relevance and meaning as a common mistake leading these changes in your organization. She defines it as "not clearly linking the change effort to the market and business strategy to create clarity in the minds of stakeholders." Ask the question, "Why?" What is the benefit of change? How does it fit within the present organizational structure? If your senior managers and leaders are not behind this change, their message to their teams will not provide the inspiration necessary to see the change fulfilled.

If your organization is run top-down, there needs to be an explanation as to how the change fits the current business strategy. For example, assume you perform sewer and grading work during road construction projects. The owner of the company wants to get into paving. You are performing the underground work; why not pave as well, so you can self-perform the work? The answer may be costly. Equipment, labor, and materials do not come at an inexpensive rate. Learning at this cost could sink the organization. Crews may become frustrated trying to learn new skills. Management may become upset because more is being added to their already full plates. No one sees the up side because the owner does not explain why this change is important to the organization. Business strategy has always been sewer, grading, and underground utilities. Hiring a paving contractor has not been an issue in the past, and you have a great working relationship with a few contractors in the area. These questions need to be answered by the owner to create buy-in from the managers and crews. If people buy in, the success rate increases. Clarity of the mission helps create buy-in.

Provoost suggests circling back to ensure that clarity remains. Setting and forgetting does not work in project management. There have to be feedback loops in place to see what is working and what needs improvement. Managers and crews asked questions of senior leaders during the change. Now, it is the senior leaders' turn to ask the questions. Remaining objective, ask managers and crews how the new change is working. Does it still make sense? What is good/bad about the change? If you do not have feedback loops, there is no way to gather accurate information about the change. You may hear rumblings through the grapevine, but those are not clear indications of the progress being made. Keep relevance and meaning at the top of your list when implementing change. The organizational strategy will change. How this change is implemented will determine the future.

Chapter 5 takes on challenges and how to exploit them. Exploit is a positive risk response. As project managers, we want to see challenges as opportunities rather than roadblocks. Freezing and thawing your challenges is a technique that allows these problems to be fully vetted. Optical illusions are discussed because it gives an insight into perspective. How we see something impacts how we act. Keep our lines of thought as open as possible. Become a problem finder instead of a problem solver. This take on project management is not often seen. It keeps the project manager proactive. I also talk about how culture and mindset become obstacles. The environment itself can be the biggest obstacle. Chapter 5 is a great mindset shift that will help you see opportunity in difficulty.

Chapter 5

Exploit Challenges

It is important to develop the skills to take on challenges. As a youngster, basketball was the sport I was naturally inclined to play. I enjoyed the game, I liked the team aspect, and I was pretty good. Had I decided to take on high schoolers, I would have lost terribly. Probably without scoring. There is a developmental process that must take place to tackle the challenges placed in front of you. In my example, you have to learn how to shoot, how to dribble, how to pass, plays to run, and so on. Without a plan or the necessary skills, you will look like a fool. Along the way, there will be times when things click. I remember playing in a game in third or fourth grade. I got the rebound, dribbled down court, stopped from about 15 feet out, and shot a jumper that was nothing but net. While that instance was rare, it was a time that sticks in my mind because all the fundamentals came together at once to produce a successful result.

The learning process remains the same as we age. Once the fundamentals have been nailed down, it is time to incorporate more advanced moves and techniques. In project management, if you do not know the basics of the profession, it will be nearly impossible to run a project in that industry. Identify the basics and fundamentals, master them, and then seek challenges to test your abilities. Some will suggest trial by fire, but that will lead mostly to frustration and an overwhelming desire to get out while you can. The goal for your team is to be pushed, but not to the brink of disaster. Be aware of your team's progress and constantly fuel the fire when you see them performing well. Purposefully take on challenges for them and see how they react, knowing they have the ability to tackle anything. Constantly train them in the fundamentals to keep them sharp. Do not assume that everyone is on the same page.

Creating obstacles can be an exercise used in developing that muscle. Remember, as a child, when the floor would be "lava," and you would have to avoid going into the lava by any means necessary. Usually balancing awkwardly on the couch or leaping to the next safe haven before adults would put an end to the game. If you feel your team needs to be challenged, create a game for them. Use it to flex their solution-seeking muscle. You want your team proactively seeking challenges, not avoiding them because of the risk of looking stupid. Lessons learned during the game can be applied to real-life scenarios. Every team-building activity I have been a part of has gone the way of silliness or people downplaying its effectiveness. These techniques work if applied appropriately with the right mindset. Get your team to see the benefits of participating rather than playing it off as some corporate time wasting.

Use the element of surprise as a tactic. No one likes pop quizzes, but they can be used as a tool to test the readiness of your team. In high school, our basketball team had a press called "Lightning." It was an all-out half-court press intended to be used once for a quick basket. Once the dribbler would cross half-court, we would send two defenders after him, while the other defenders covered the nearest opponents. This defense would always leave a man wide open underneath the basket. However, if the team did not expect it, he would pass to the nearest teammate, and one of our players would steal the ball for a layup on the other end. If they were prepared for the call, he would throw it to the wide-open teammate under the basket for a layup. Having a drastic contingency like this may be necessary for some projects or teams. Prepare your team for it, so when it has to be used, it can be deployed with precise execution. The element of surprise for your team or the stakeholders, if used properly, can be an effective strategy for overcoming obstacles. It may be using a different material for a product or going to a different supplier for the same material. Shaking things up a bit, when timed correctly, opens the eyes of everyone and refocuses attention.

Be proactive. It is crucial for you as a project manager to remain proactive rather than reactive. Being reactive means it is too late. The issue is over, and now you are trying to find a way to solve it. Remaining proactive means you are seeking out potential issues and resolving them before they happen. In the construction field, project managers have to envision the project on a physical level, not a planned view. Everything always works out on the paper plan. The utilities are not in the way of each other. The sewer system connects easily. The grades provided do not interfere with any of the installed sewer systems. In reality, this is not the case. Utilities are almost always in the way. Aggregate and dirt quantities need to be kept track of mentally, so you do not take too much dirt off the project or bring too little aggregate on to the project. It does not have to be an exact science. This approach is where a project manager's estimating ability comes into play beyond bidding the work. If you have 1,000 feet of road to grade at a width of 25 feet and 6 inches of stone is required, you have approximately 925 tons of stone to be placed. If each truck contains 22 tons, there are going to be 43 truck loads coming onto the project for that area. Assume each truck has a half-hour round-trip time. For the entire day, you want to have three trucks on the haul to take up the day. These are the kind of constant calculations you need to be making as a project manager. Sure, it would be nice to have ten trucks on, but the job would take half a day, and the resources you have could be used more efficiently. On the flip side, one truck is going to extend the task by a full day, potentially causing delays. The calculations take five minutes and can save you a day's work. This mental math is how you become proactive. Seek out where things can go wrong and do the math to make them go right. Per the example, if you find yourself being short on material (reactive), it will be nearly impossible to find a truck midway through the day. Pass these mental checks onto your team as well. Do not assume they know this or that they are doing this already. Constantly remind them to check. Follow up with their progress. Get them performing. Otherwise, you will be doing more work than necessary and reacting to issues that should have been easily solved.

5.1. Stoicism

The impediment to action advances the action. What stands in the way becomes the way.

– Marcus Aurelius, *Meditations*

In this chapter, we are focusing on becoming problem-seeking and solutions-oriented. Find the issue. Solve the issue. The quote ties directly in with our new mindset. If there is something in the way of progress, that something is where you should be. Do not look for others to solve it. Ask for help, but do not rely on others for solutions. Use your creativity and ingenuity to come up with an answer. Many of us played video games as children. How many times did you come across a path that was blocked or needed a secret key to open? Was that always the path you needed to go down to advance in the game? That is exactly how you should attack the problems in your way today. Many times, it is simply asking where the key is. Sometimes you will have to ask someone who will lead you in the direction where the key is, but you will still have to search for it. Once you find it, then you will have to return to the blocked path to open the door and continue on your mission. Here is where the road is not necessarily straight and narrow. It has winds and bends and hills and valleys and so on. You will not always possess the skills or knowledge to unlock the pathway. You will have to rely on others to help you find the key to unlocking the door.

In road construction, there are always snags along the way. In building a road, it seems logical to start at one end and finish at the other. But sometimes that is not the right way. There may be emergency accesses that need to be put in before the project can really move forward. Businesses need to stay open. People need to get to work. All of these issues can cause a project to take a different turn than originally planned. Each of these issues needs to be dealt with before the project can continue. These issues are the impediment to action that Marcus Aurelius refers to above. Having these setbacks lets you know you are progressing in the project, because more keeps happening. Cause and effect. If you are at a standstill, no progress is being made until the challenge has been met. Continuing to run into roadblocks (pun intended) lets the project manager know that progress is occurring. Do not run from these challenges. Take them seriously. Doing so leads to more progress, more production, and more revenue. Proactively seeking out problems is the mindset goal for you and the team. Anticipate an issue coming up, and have the solution ready.

Most road projects have a cut/fill aspect to them, meaning dirt from one area of the project can be used in a different area. The goal is to use as much of the on-site dirt and aggregate as possible, to limit expenses for hauling material to the site. If the project manager is not taking a mental inventory throughout the day, an issue could arise from having too much or too little material on-site. This is where the problem-seeking mindset thinks proactively, having a solution for too much material (an area to stockpile or a dump site close to the project) or too little material (haul from a source close to the project). Even if the excess material comes as a surprise that was not anticipated at the planning stage, senior management wants to know where the material can go and that you are thinking far enough ahead. When a stage of the project approaches, and you have too much material and nowhere to go with it, that is not a good look. Get your team to search out possible issues they may face and have solutions ready to use. Do not rely on your ability to track all of this information. Things will be forgotten and overlooked, causing delays or further hits to the budget.

5.2. Taoism

> *The bright Way appears to be dark;*
> *The Way that goes forward appears to retreat;*
> *The smooth Way appears to be uneven;*
> *The simplest reality appears to change.*
>
> – Lao Tzu, *Te-Tao Ching*

Not only Stoic philosophers decided the difficult path is the path of choice, but also the Taoist philosophy held the same notion. Again, the mindset may seem counterintuitive at first. People tend to be like electricity, taking the path of least resistance. In doing so, though, people do not reach their full potential.

They get comfortable, content, satisfied, and so on. Do not become one of these individuals. We are striving for greatness—not just meeting expectations, but exceeding them. Think of someone who appears to have life easy. Everything this person does is a success. Everyone they are around is beautiful. Places they go are out of a magazine. But is that "smooth" path really that smooth? Behind every great success story is a work ethic second to none. Unless someone is truly "lucky," hard work and dedication are required to reach that level of success. When watching a sporting event, how often do you think of the work those players put in before appearing on national television? I rarely think about it. Then I will see a video on the Internet of the countless hours put into training not just their bodies but their minds as well. Changing the mindset can have a greater impact than changing the physical body. All professional athletes have natural abilities and superior conditioning that allowed them to reach that level. The difference is between the ears: the mental game. Not giving the moment too much space to bounce around. Treating each game as a game. Doing the things necessary in training to limit mistakes when fatigue occurs during a game.

That is where we want our team members to get. Train them to deal with high-stakes situations before they occur. When someone cuts through a gas service on a road project, we do not want everyone standing there looking at each other, waiting for someone else to do something. We want each of them to understand that making the call to the utility company is a priority. Constantly discuss safety on the project. Every morning, go through who needs to be called for each circumstance. While major problems may be few and far between, it is when they do happen that quick response counts the most. Talking about it once but then never talking about it again until it happens does not get the point across. Continuously discuss expectations for the team. Give the foremen daily goals to reach when undergoing a task. If we bid the job for 16 feet of concrete pipe to be installed every hour, tell them we need 20 feet. Expect more out of them than they do of themselves. Set a performance standard so high that failing is not an option.

Reward individuals who actively seek out problems. Encourage that behavior. Become an example of someone who practices the same approach. Do not delegate the issue because that is the easiest thing to do. Give someone the opportunity to solve the issue because that is where the person will succeed. If you are the best for the job, take on the issue and show your team what you expect.

5.3. Sales

In this section, I want to discuss the power of committing to an idea. It is going to take massive amounts of commitment to overcome obstacles. People who are not committed give up easily. They do not have the intention of finishing what they started. This mindset will never accomplish the success we seek. You need to be committed to what you are selling, otherwise the result will be "No's" across the board because customers will feel your lackluster passion for the product or idea. However, if you are committed, you will be able to convince customers they not only want what you have, they need it. Commitment to a project can be difficult to establish, especially if team members have been down a similar road before. How many times have you been a part of something you just know will not work but have been assigned regardless? What kind of effort are you going to put into that project? Minimal at best. Lack of commitment will lead to people avoiding challenges and trying to take the path of least resistance. With full commitment, team members are going to attack anything and everything in their way. They will be proactive in their search for solutions. Sell others within the team as to why this idea is going to work. Create the buy-in we seek.

Think of the scenario of you and your team on a ship setting sail for an island. Once you leave the shore, it is all up to the individuals involved as to who survives, if anyone. Success is getting to your destination. It does not matter the route taken. It does not matter how the ship looks when it gets there. It is all about reaching the destination with everyone on board. A project is the same. You and your team are in on this together. You have to sell them commitment to the idea. Sell them on the destination you want them to reach. Make them aware of the opportunities that lie ahead. Once you have that commitment, the ship will nearly steer itself. You will be the captain sitting back and looking at your team members doing what you set out to do in your plan. Selling an idea to which you are committed is easy. Now it is time to get your team on the same page you are. Ask questions to create commitment. If the conversation is one-way, people tend to block you out at a certain point. Get that feedback loop working strong. Make your team's voice heard. Show them your commitment.

5.4. Freeze-Thaw Your Challenges

A freeze-thaw cycle causes havoc on roadways. The summer construction season is dedicated to repairing roads damaged by a heavy freeze-thaw cycle. Imagine doing the same thing but with ideas your team brainstorms. A freeze-thaw process to expand the idea to utopic regions and contract the idea sure to fail. Push the boundaries at both ends of the spectrum to hash out each possibility. Take your ideas to the extreme and then find common ground of actual capability. The freeze-thawing of ideas will allow you and your team members to expand and contract mentally. Maximize their mental capacities for thought and vision.

Think of your brain as the roadway during this cycle. It will be tormented with 100-degree heat for months, then bombarded with 20-degree cold for more months. If that roadway holds up in those conditions, it is one strong road. Now imagine your mind going through a similar process. The outcome will be a strong mind capable of thoughts far beyond its initial capacity. You will be tested and you will triumph. The cycle will run in reverse. It will be a thaw and then a freeze. You want to expound the challenge as far out as you can. Heat the issue up with as many ideas and outcomes as you can imagine. Even if they are outside the realm of possibility, write them down. Now that you have all the possibilities laid out, it is time to freeze the process. Shrink it down to the most minuscule of details. This cycle will lead to breakthroughs along with other ideas to expand the organization.

Once you have the process frozen to a polar-like state, start to play with the words of your challenge. Use synonyms to further define your talking points. Antonyms can also be used, so each team member knows the direction you do *not* want the team going. Ask "Why?" and "How?" until you cannot anymore, then do it one more time to make sure. Question the problem differently. Use opposites to flip the script. Instead of asking "Why?" start to ask "Why not?" Team members will begin to start thinking on both sides of the fence, hashing out each challenge from all perspectives. Replace negative words with positive ones. This cycle is all about getting your team in the mindset of getting things done. See challenges as opportunities. Embrace the change to take on difficult tasks and resolve them. Negative words bog a team down, even if they are written down for only a short period. When things are rolling, and momentum is going your way, do not stop it by bringing up negativity. Continue to let the ideas flow, and once there is a down period, be the devil's advocate. When an issue arose and the question, "Can we do it?" was asked, a senior manager of mine would always retort, "We always do." That was light in a dark situation. He would always inject hope, no matter the issue. That makes a difference. If the leader thinks it can be done, the team members will be close behind with their support.

5.5. What Do You See?

Once we have settled on a perspective, we close off all but one line of thought.

– Michael Michalko, *Thinkertoys*

Optical illusions have long been of interest to me. One can stare at a still image, and it appears to be moving. I can look at an image and see a duck, while another person looks at it and sees a rabbit. In this chapter, we have discussed many ways

Figure 5.1   Line illusion. (*Source:* www.brainden.com)

to look at a problem from many different perspectives. Optical illusions present a physical way of doing the same thing we do mentally when approaching a problem. In this section, there are five examples of classic optical illusions (all the images can be found at brainden.com). These show how your perspective can be deceived, and how changing your perspective can create an entirely new image. Whether it is staring at the image in front of you or mentally at the challenge ahead, perspective has a heavy influence on our ways of thinking.

An initial glance at Figure 5.1 may show bent lines. If you look more carefully, then you can see that the lines are actually parallel. In a previous chapter, I mentioned how organizations can overlook risk analysis, not only glossing past potential risk but also creating a risk that is not present.

Figure 5.2 shows two identical tables realigned to give the illusion of two different dimensions. Again, changing your perspective on a problem may show the

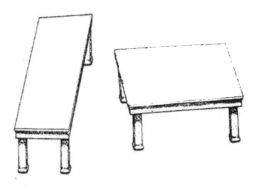

Figure 5.2   Size visual illusion (tables). (*Source:* www.brainden.com)

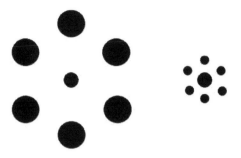

Figure 5.3 Size visual illusion (dots). (*Source:* www.brainden.com)

answer. Looking directly at the challenge ahead may disguise potential opportunity not seen by others. Turn the challenge slightly to uncover the benefits.

Titchener circles (Figure 5.3) is a classic optical illusion. The illusion is that the size of the central circle in the two images appears to be different. In fact, the two central circles are the same size. If you are creating a physical product, spatial awareness is important. Creating the illusion of a smaller object by surrounding it with larger objects may fulfill the requirement far better than redesigning an object to be smaller. Changing the owner's perspective may be the answer, rather than changing the team's perspective.

Old lady or young woman in Figure 5.4? The mouth of the old lady is the necklace of the young woman. The nose of the old lady is the chin on the young

Figure 5.4 Face illusion. (*Source:* www.brainden.com)

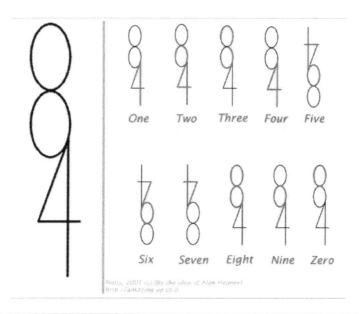

Figure 5.5 Numbers illusion. [*Source:* Natis, 2001 © (By the idea of Alon Heimerj). Accessed August 3, 2016, from www.brainden.com.]

woman. These images are meant to be looked at with a quick glance to gain first impressions. There is no wrong answer, but it will give answers as to how you are looking at things and whether you see the entire picture.

Figure 5.5 illustrates how each single-digit number can be created by combining two numbers together. Often, project managers have difficulties due to a lack of resources. This illusion is meant to give you an idea of how one resource can be used many different ways. Take a shovel, for instance. Its main purpose is to move material from one location to another. However, it can also be used to hammer in stakes, as extra leverage to pry materials apart, or even help a person climb a steep grade. Seeing a shovel as just a material mover will limit the number of ways it can be used. This principle can be applied to many other resources, including personnel.

"What do you see?" is meant to keep your mind open and free of confirmation bias. As the quote states, once you pick a lane, rarely do you change lanes. Not only mentally, but physically and professionally. Career change is seen as a bold move. After not working out for decades, starting to diet and exercise comes off as extreme. Doing something the same way for many years and then deciding to switch to a new approach throws off everyone around you. As an entrepreneurial project manager, this is where you want to be: unpredictable in your ways. Try new approaches and principles. Challenge your team to new heights. Change your perspective to see the opportunities that lie ahead. Instead of focusing on the risks, present new ways to turn those risks into opportunities. Where there is a great risk, there is a great reward.

5.6. Problem Finders vs. Problem Solvers

"Knowing what causes something is the first step in preventing it from going into effect," wrote Jim Paul in *What I Learned Losing a Million Dollars* (Paul and Moynihan, 2013). Finding the problem to fix can be more difficult than solving the problem. Once the problem is identified, the resolution becomes apparent. Here lies the difference between problem finders and problem solvers. Some can identify an issue and find the quirks, while others have the ability to solve the problem at hand. Combining those skills supercharges a team's ability to succeed.

In *Little Bets* (Sims, 2011), Peter Sims writes of psychologists Jacob Getzels and Mihaly Csikszentmihalyi conducting a study to spotlight the importance of problem finding to creative work. Getzels and Csikszentmihalyi showed 27 objects to 35 artists, asking them to draw using some of the objects. Artists labeled as "problem finders" took the time to study each of the objects and decide how best to incorporate them into a work of art. They saw more possibilities and had the ability to change on the fly if new opportunities presented themselves. Artists labeled as "problem solvers" immediately started to draw using objects they saw first. Instead of finding new and interesting ways to incorporate the objects, they took to getting the job done and fulfilling the requirements. Independent judges found the work of "problem finders" to be vastly more creative than the work of "problem solvers."

As discussed with leaders being heralded over followers, problem solvers get credit, whereas problem finders do not get as much recognition. Putting out fires falls in line with management practices more than taking a proactive approach to finding the potential issues before they occur. Switch your mindset to problem finding rather than solving. How can this be done? Here are some ways to become a problem finder:

- *Turn presented problems into discovered problems.* "If I had asked people what they wanted, they would have said faster horses," said Henry Ford. Presented problems are already somewhat defined; they are waiting for you to solve them. These are your daily fires to extinguish. Discovered problems take curiosity, research, and an internal dialogue to help construct and define the problem.
- *Ask better questions.* Simply asking "Why?" over and over can get you better answers. The problem rarely lies within the first answer. Asking the first question without following up will not get you to the root cause. You

must keep digging. It will be uncomfortable, but the treasure trove of information that awaits is worth it.

- *Have an internal dialogue.* Have continued discussions with yourself. Jeff Fajans, of Create.Learn.Live, writes "How to Become a Creative Badass" (Fajans, 2014), suggesting asking yourself these questions:
  - ○ "Is what I am doing working?"
  - ○ "How and in what ways can I make improvements?"
  - ○ "Am I even working towards the right goals?"
  - ○ "Am I excited about the direction I am taking?"

Problem discovery, in part, deals with self-discovery. Instead of blaming the issues on subcontractors and other stakeholders, take a look at what you are doing. Always try to get better. Discovering problems leads to greater heights. Solving problems leads to a feeling of productivity but, in reality, you are only spinning the hamster wheel.

5.7. Becoming an Insider

We are outsiders at the beginning of any journey we take. There are nuances and experiences we must obtain before reaching the next level. You sit back at meetings and take copious notes to learn as much as you can. You listen to everything for any nuggets of information the conversation provides. You speak when someone asks you a question. Your role as the outsider is to become knowledgeable and do any work, even if it is completely boring and is something no one wants to do. At some point, you want to become an insider. One who stands tall with confidence in meetings. One who is respected when speaking and is someone who takes on a leadership role remembering those followership days like yesterday.

Playing video games is an activity most people have experienced. In the beginning, you are button mashing—pressing every button on the arcade console or controller to find out which button to push. You may have some success with this technique, but never consistently. One day, someone comes along and turns your button combinations to special attacks. If you hit these buttons in succession, a special attack move will be performed that causes more damage than your button-mashing technique. You are gaining insider status. Then, you pick up a magazine with more tips and tricks inside, showing you where the opponent's weak spots are and letting you hone your attack skills. The level boss could be most vulnerable when the boss's mouth opens, or using the boss's henchmen to attack could cause more damage than you could. As the game goes on, your weapons become stronger. Instead of a tiny bullet shooter, you have access to a rocket launcher. You become full insider when you discover secret hacks or "Easter eggs" (glitches within the game). All of this takes time and appreciation for the game.

The same goes for project management. In the beginning, you are a button masher trying every tactic you learned in school or in the field. You call the boss anytime something goes wrong. You do not have the wherewithal to handle conflict on your own. Once the boss sternly tells you to figure it out, you start to develop your strategies for dealing with conflict. Your insider is developing. Eventually, you start to take leads on small tasks, gaining confidence to do more. You get outside of your own head. Instead of visualizing all that could go right or wrong, you start actualizing. You do the work and find out the consequences, good or bad. Outsiders overthink and start to worry without anything even happening. They are given a task and immediately see it failing. Insiders know the task, have a system in place for getting it done, and if something should go wrong, deal with it then. As in the video game example, your weapons become sharper and more honed. You are no longer afraid of the unknown because you have the tools to deal with it. Insiders get excited about the next level or challenge. They want to put their skills to the test.

5.8. Five Strategies to Overcome Obstacles

As this chapter has discussed, obstacles and challenges come your way constantly. It may be by the minute for some, for others by the hour, and for most by the day. Robert Greene, the author of Mastery (Greene, 2012), developed five strategies to deal with the main obstacles in your path over time. The strategies are as follows:

1. "The primal inclination strategy—a return to your origins. For masters, their inclination often presents itself to them with remarkable clarity in childhood.
2. The Darwinian strategy—occupy the perfect niche. You see others there making a living and treading the familiar path. The game you want to play is different: to instead find a niche in the ecology that you can dominate.
3. The rebellion strategy—avoid the false path. A false path in life is something we are attached to for the wrong reason (money, fame, attention, and so on).
4. The adaptation strategy—let go of the past. In dealing with your career and its inevitable changes, you must think in the following way: You are not tied to a particular position; your loyalty is not to a career or a company. You are committed to your Life's Task, to giving it full expression.

5. The life-or-death strategy—find your way back. No good can ever come from deviating from the path that you were destined to follow."

These strategies can work on the project level, organization level, or career level. The primal inclination strategy is focused on getting back to basics. Sometimes project managers overcomplicate things by getting inside their heads. The solution can become obvious by taking a child's perspective. Go back to being an outsider, mentally trying to gain the most knowledge possible and sitting back listening. The Darwinian strategy is most likely applied at the organizational level, trying to find the next business venture, selecting which projects fit best in the program or the best programs for the portfolio, taking a step back, surveying the land, and finding the nooks and crannies to specialize in. The rebellion strategy is good for all the levels mentioned. Are you in the right organization for you? Is this project the right one for the organization? Are we going about this project in the right way? This strategy can be used as a check and balance for the direction you, your project, and your organization are headed. The adaptation strategy is about you and your career. Are you defining yourself by what you do? Does your career give you satisfaction, or is it a mean to other ends? It is an introspective strategy for which only you have the answers. The life-or-death strategy brings you back to the question "Why?" What is all of this for? Have you become so involved with your career that the most important things in life are taking a back seat? Another introspective strategy is to find out what path you are on and the reasons behind it.

5.9. Obstacle: Culture

The biggest obstacle you face may be the culture of your organization. I will define culture as beliefs and behaviors of an organization, normally implied rather than defined. The implication of culture can make it difficult to determine how to proceed when an opportunity presents itself. Does it need to be the idea of upper management to proceed? A stalemate can occur when the team believes in an idea, yet upper management is not sold. Change is not easy for a large organization. The difficulty compounds itself when the culture is not ready for a change. Sandrine Provoost delivered a webinar titled "Preventing the 10 Common Mistakes in Leading Your Change and Transformation Projects" (Provoost, 2016). To assess your organization's culture, she wants you to ask the following questions:

- What already supports your future state?
- What directly block it/needs to be dismantled?
- What needs to be created for you to succeed?

These questions are geared toward a company leading change transformation projects, but you can also ask them for all projects. How can you get your organization to help you instead of inhibiting you?

If culture becomes a major obstacle in your projects, moving forward, you must include that culture in all preplanning efforts. Working with it rather than against it may prove easier. Culture may be the immovable object in your planning. Working with culture creates a momentum in your favor that cannot be stopped and is instead immovable. Find the spaces in your organization's culture that support your ideas and gear your projects toward the support. Changing organizational culture will prove to be insurmountable. You will waste time and energy that would have better served your projects. Testing those aspects of culture will be a new task for you as a project manager, not only estimating which aspects work for you but also knowing which work you want to do. In this case, knowledge is power. Testing, failing, and succeeding will all lead to knowing.

5.10. Obstacle: Mindset

Organizational culture will be an obstacle, but so will your mindset. "Mindset is to the individual as culture is to the organization," says Provoost (Provoost, 2016). Your mindset is important, but more important are the mindsets of your organization's leaders. Who are the leaders? Titles and leadership do not go hand in hand. Some bosses are partial to others and listen to them more. How do they think, then act? Get a sense of how the leaders think about issues and come to solutions. How much data do they need for decisions? Are they intuitive or analytical? Find out how they make decisions and use that to your advantage. Provoost gives us three mindset change strategies:

- Get agreement first
- One-on-one coaching
- Team development on group mindsets; agree on changes and how to monitor this personal work over time

Start with getting agreement. Does your team, stakeholders, or organizational leaders want to change? If not, forget any future change plans you may have. They have to want to change to change. Seems elementary, but can get lost in big initiatives. Once they agree to change, get them one-on-one coaching to get to the cause of the issues. Experts can see things we cannot. Sometimes we are so close to the situation it blinds us. A subject-matter expert will get a plan working for these individuals. Now they have agreed to change, have a plan from an expert to change, and then start to develop team mindsets to get everyone on the same page. Once they are in the performing stage with new mindsets, success starts to come. You will find this improved success rate by monitoring their work over time. Getting quantifiable data to measure against works at the organizational, project, and personal levels. Establish a set of metrics, produce a spreadsheet, and then begin analyzing. Find what works, present this information, and continue to thrive.

Chapter 6 takes a different approach to agile management. Agile project management is becoming more accepted. The change to a new way of management has proven difficult. I discuss the S.C.A.M.P.E.R. method developed by Michael Michalko. This method will take your sprint sessions to the next level. It is a 360-degree view of the problems at hand. I also take a look at to-do lists. A project manager has task after task listed as to-do items. The list can go on forever. This new viewpoint gives an updated look at making sure what is on the list needs to get done. If you are new to the agile setting, Chapter 6 discusses ways to develop as an agile leader and what to look for. A survey done by Lawrence Cooper asked executives and senior leaders their thoughts on agile management. Their answers provide great insight into the agile world.

Chapter 6

Entrepreneurial Agile Management

In this chapter, I will speak of agile management in the organizational setting. It will apply at the team level, but the main focus is on the organizational level. This chapter will tackle new ways to approach challenges and how failing quickly helps your organization. An emphasis in agile management is to see what is working and what can be removed before the project becomes too established. I will also speak about how to reorganize your priorities to ensure they fit the organizational model and outlook. Do not allow your project management muscles to atrophy. Employing agile management allows a project manager to flex that muscle and remain in the game constantly. It goes back to creating unease and resisting that comfort level. Every sprint session should present new challenges to tackle and new skills to obtain. Use these not only to make the product or service better, but also to make yourself and the team better. Constantly come away with lessons learned. Never go a full meeting without a teaching moment. Keep your team informed.

For complex projects, agile is the preferred method. It breaks down projects into smaller, more manageable chunks. In construction, staging is used to break down a project. Instead of focusing your efforts on the entire project, start with the utility work, move to the storm sewer and water main, grade the road, and then finally pave. Reducing complexity should be a constant focus for the project manager. Whether it is at the task level or the project level, the simpler, the better. Later in this book, the 80/20 principle will discuss this in depth. Get results from the fewest moving parts.

6.1. S.C.A.M.P.E.R.

Michael Michalko, the author of *Thinkertoys* (Michalko, 2006), recommends S.C.A.M.P.E.R. as a technique to brainstorm new ideas from existing frameworks. S.C.A.M.P.E.R. is an acronym for Substitute, Combine, Adapt, Modify, Put to other uses, Eliminate or minimize, and Rearrange. During the sprint sessions of an agile project, use this template to improve upon the deliverable. It is a way to over-deliver on a project without abusing the triple constraint (scope, schedule, and cost). This technique forces you to think about the product or project in different ways by asking many questions at each stage of the challenge. Each time your team meets, whether it is every day or week, every two weeks, or once a month, these questions should be asked to ensure you are taking on the challenge from every angle. Later, I will discuss each letter of the acronym and its importance to the process of creating the best product or service for your customer.

S.C.A.M.P.E.R. applies not only to the product you are developing but also to the process by which you are creating the product. Each of us likes to think we are using best practices. As you have read, thinking you are using best practices is putting you behind the 8-ball, because processes and practices should always be improved. This technique is all about spinning your reality on its axis to create something better from the resources you have. Constant improvement is the name of the game, and S.C.A.M.P.E.R. allows you to stay ahead of the curve.

Consider a scenario in which we can use S.C.A.M.P.E.R. My project is reconstructing the roadways of a subdivision. The aspects of the project include storm sewer, watermain, sanitary sewer, excavating, grading, paving, and landscaping. This technique will work at each of the aspects, but for now we will focus on grading. Grading the roadway involves removal of the pavement, placing of the stone, and grading of the roadway to finish elevations. As Michalko lists, the following are questions to ask at each stage of S.C.A.M.P.E.R.:

- What procedure can I *substitute* for my current one?
- How can I *combine* grading with some other procedure?
- What can I *adapt* or copy from someone else's grading methods?
- How can I *modify* or alter the way I grade?
- How can I *put* my grading *to other uses*?
- What can I *eliminate* from the way I grade?
- What *rearrangement* of grading procedures might be better?

In the next sections, these questions will be answered.

6.1.1. Substitute

Pavement removal can be done in many ways. A backhoe can be used to rip up the pavement as is, without using a jackhammer or stomper. Asphaltic pavement is the preferred material for this method. If the roadway is concrete, a jackhammer or stomper is needed to break through the material. A jackhammer attachment for the backhoe will be used in most cases, but there are different ways pavement removal can be accomplished.

Placing stone can be done in a variety of ways as well, including materials used for the roadway base aggregate. On some projects, the specifications allow the removed pavement to be crushed down into usable material on the site, saving the contractor trucking dollars. If this is not the case, a material supplier must be found in the area, to reduce those same trucking dollars. Also, how you place the aggregate may vary. On larger projects, a bulldozer is used to rough grade the road. Rough grading the road means getting the material within inches of the required depths. After the dozer performs the rough grading, a grader will perform finish grading to the exact elevations in the plans. Smaller areas may require a skid steer and hand work, i.e., shoveling, to get the aggregate within specification. If you can replace a bulldozer with a skid steer, your costs decrease substantially.

Grading the roadway to plan elevations can be done using two techniques, rotary laser and Global Positioning System (GPS). Smaller projects can be done using both techniques while staying effective. Larger projects should be done using GPS. A rotary laser and grade rod require a skilled laborer to benchmark each time the laser is moved, perform calculations on a regular basis, and be aware of all operators at all times. This task is complicated and results in mistakes that can be avoided with GPS units on the machines. If your organization can substitute the rotary laser with GPS, you will save time and money in the long term. The GPS units connect to the machine performing the work, and the plan grades are uploaded. The operator keys in the stations of the day's work while the machine does the thinking. GPS is more accurate and does not require the skilled laborer to follow the crew.

Substitution on a construction site may be the easiest and most cost-effective strategy of the S.C.A.M.P.E.R. method. As long as the project meets specifications and is built to plan, the way to go about it is up to the contractor. If you can get the job done with a skid steer and a laborer, by all means, go for it.

6.1.2. Combine

The simplest way to combine on a construction project is to fast-track the schedule. For instance, the utility installers can be working at the same time the road is being graded. Even the utility installers can perform multiple installations at once. Storm sewer and watermain can be installed simultaneously, as they are sometimes at similar depths. Combine your tasks to perform the same work in a shorter amount of time.

Combining skill sets is an advantage as well. If you have an operator who can perform laborer tasks, you are afforded the benefit of killing two birds with one stone. Instead of hiring more people or stealing from Peter to pay Paul, you have the ability to assign more tasks to one individual. Look for ways to stretch your talent pool. Give them opportunities to prove themselves. This works the other way as well. Find a laborer who can give you some operator skills. Usually, a laborer can perform simple operator skills such as running a skid steer. This way you receive the benefit of a dual-threat employee. The more dual threats you have, the more work can be performed.

6.1.3. Adapt

Many times you will be performing work with other contractors. You will have to deal with their ways of doing things, and vice versa. Do not forget to be observant at those times. Look to see how they perform their work; if you learn something from it, adapt their way. For example, building mechanically stabilized earth (MSE) walls on highways projects requires tying panels together with wire. This installation can be done a multitude of ways. The first way a contractor tried was doing this by hand, having a laborer thread the wire through the panels and use pliers to twist them together. As you may note, this took longer than expected, and the contractor had to adjust; otherwise, every day would be performed at a loss. After making phone calls and doing research, the contractor found a tool that tied the wire instead of having the laborer twist it with pliers. After purchasing the tool, the panels started flying up, and production increased immensely.

After work is performed, getting paid the proper amounts can be a hassle. Measuring depths, lengths, and widths by hand is not the most accurate of ways to reconcile, but using surveying equipment while performing the work changes the game for the contractor. While a contractor grades out the road, use the survey equipment to pick points that will accurately depict the area being constructed. Once work is done for the day, take those measurements and calculate quantities. Document these quantities daily, so if there is an argument after the work is performed, you have the data and supporting documentation to defend yourself. Adapting to technology can be difficult initially but often pays off (literally) in the long run.

6.1.4. Modify

Because you have performed work one way forever does not mean adjustments should not be made. Go back to the rotary laser and GPS situation. Why switch to GPS if you have built roads for 30 years using a rotary laser and grade rod? If it ain't broke, don't fix it. Taking a different approach may seem scary at first, but becoming an entrepreneurial project manager is going to be a scary road. Instead of seeing technology as a young person's game or looking at the high initial cost of it, start to break down the return on investment. It may seem obvious, but I have seen so many opportunities go to waste because someone will not put in the time to research the benefits. They see only thousands of dollars being spent and immediately discard the idea. I have seen GPS work wonderfully for so many contractors, yet I continue to hear push-back on making the switch. If you want to play with the larger companies, you have to emulate them. Modify your approach. See how they do it. Emulate it to fit your budget and organization.

6.1.5. Put to Other Uses

Grading a roadway seems straightforward. What other uses could it serve? Grading the roadway ahead of the underground utility installers is a prime example. The grading crew can help in the excavation of the underground utilities, especially if a large cut is required. All grading projects have a certain amount of cut and fill to the project, "cut" meaning taking material away and "fill" meaning bringing material onto the site. If there is a large cut in an area of the project where one of the utilities needs to be located, the grading crew can rough grade that area, so the utility installers will have less material to excavate during their work. So not only is the grading crew getting the road to plan grade, they also involve themselves with the utility installation to make everyone's life easier.

This process can also work in reverse. The utility installers can help the grading crew by not leaving piles of excavated dirt laying around and bringing their area to rough grade. During the installation of storm sewer, watermain, and sanitary sewer, the excavated dirt is replaced with a more stable material, usually stone. The excavated dirt needs to be hauled away, either off site or to an area of the project that will accept fill. If neither occurs, piles of unwanted dirt can stack up around areas that eventually need to be graded completely. If the utility installers do not clean up their area, the grading crew now has more work ahead of them. The utility installers can become part of the grading crew at this point, remove their excess materials, and shape up the area so the grading crew can proceed full steam ahead.

6.1.6. Eliminate or Minimize

The three main aspects of construction that need continued observation and evaluation are labor, equipment, and materials. Eliminate or minimize will be most obvious in those three areas. Take a look at labor. If you bid the project for four laborers, but you see the need for two, then eliminate two from this project and move them to a project suited for more laborers. Seems obvious, but again, I have seen many projects with laborers standing around or looking busy because there is not any work for them. You are taking on extra costs for the sake of saving face.

Equipment is next. I worked for a company that charged each division a day rate for their equipment. No matter how many hours that piece ran, you were charged for the full day. For a project manager, that emphasizes the need for the equipment. If you do not need it for a full day, how can you best use the piece of equipment so the day charge does not impact your bottom line more than it should? Maybe the skid steer needs to be there all day because tasks pop up throughout the day that require its presence. A skid steer is great for handling materials. There is less risk of injury to a laborer, plus it is more effective in carrying heavy loads quickly and over longer distances. If equipment does have to sit, see if another crew can use it for that day. Larger organizations have multiple projects taking place within a small radius. In this case, your project does not take the hit for having the equipment, and another crew can be more productive. Projects rarely take place in isolation, meaning your resources are impacting other projects. If one of those resources becomes free, others should be able to use it.

Materials are the last of the three main components. While you probably will not be able to eliminate materials, minimizing their use or application can be possible. Trucking in stone versus crushing the removed pavement on the site is an example of minimizing the impact on the project's budget. Trucking of material, either on or off site, is where most of the cost of a road project lies. If you can keep most of your material on site and recycle it for the finished product, you are ahead of the game. Minimize the amount of material being trucked on site and possibly eliminate if all goes well. Continuous calculations are needed to make sure there is enough material on site to build the project. Do not rely on the estimates done before work starts. Mental math is your friend. Trucking too much material on the site is an easy way to get a phone call you do not want to answer.

6.1.7. Rearrange

There are many ways to skin a cat. The same goes for building a road. Which end do you start from? How are you going to remove the pavement? Are you going to remove all of the pavement first, or remove as you go? Removing all of the pavement puts you at risk of Mother Nature. If the rains come and damage the soil, resulting in taking out more material than originally planned, the excess material will not be paid for because you caused the damage. Traffic control will determine at which end you start. Rarely is the road completely shut down to give the contractor free rein. If there is a complete road block, this allows the contractor to arrange the staging of the project however it wants. The plan must be approved by the owner.

Placing the aggregates on the project can occur in multiple stages or in one large grouping, depending on the project settings. If there are cul-de-sacs, those may be paved before removals even begin on another portion of the project. Larger sections are done in sequential order, but the smaller offshoots of the project may be completed before the larger sections even begin. These can be rearranged so the approach best suits the neighborhood and the owner. Again, the aggregates can be rearranged as well. One area might suit crushing the material and recycling it better than another area. Trucking in the material might make more sense, depending on the lay of the land.

Grading the roadway is another step that can be rearranged. For staging equipment on site, moving around might make more sense because the areas to stage may be small if the project is located in a neighborhood. If it is a larger project, the mainline should be your focus, and the side streets can be worked in as you go. Rain can interfere with your plan, and working in the mud is nearly impossible. Your focus can shift to other areas of the project where aggregate has been placed. Placing aggregate is a great rain-day activity, as the rain helps the material compact.

S.C.A.M.P.E.R. is intended to be used constantly as a way to update your best practices. Every two weeks at your sprint sessions is good, but daily checks using these quick techniques is best. This technique can be written down or performed mentally, whichever is your preference. Use it as a tool for constant improvement. Doing it once and moving forward does not do any good. Give these tools to your team members as well. The more eyeballs are on the project, and the more each of you is thinking alike, the better are the chances of success.

6.2. Stoicism

The world is maintained by change—in the elements and in the things they compose.

– Marcus Aurelius, *Meditations*

Project change is an important reason to use agile as a method of management. Change is inevitable. A college professor of mine would always say, "A project is behind schedule the day it starts." If there were no changes, that would not be the case. It is a cynical view on scheduling but almost always true. One reason for change in the construction industry is soil conditions. Even if the owner does soil borings and tests, conditions always seem to change. Where soil conditions should be stable (according to borings), they turn to mud and need to be undercut even further. Think of the time added if that were to continue over miles of roadwork. That change is unexpected and can push a schedule back weeks if not months. This change is where agile is useful. Have the weekly update meetings involve the owner and contractors to make them aware of the changing conditions.

6.3. Taoism

Great completion seems incomplete;
Yet its usefulness is never exhausted.

– Lao Tzu, *Te-Tao Ching*

Think about how at each sprint session it seems there is more work than before. While you are completing work, there also seems to be an incompletion, yet you are meeting the requirements of the sponsor. The usefulness of what you are creating is never more apparent as the project proceeds, but if you let it run wild, there will always be more to add. Think of a mobile application that begins to morph as time passes. The sponsor starts to see how similar applications work and continues to want more features. Scope creep becomes a problem. There comes a time when the project manager has to say "Good enough." If the features meet the requirements, you satisfy the project's needs and goals. A point of diminishing returns is reached. The more you play with ideas, the more complicated and involved the project becomes. You are turning a molehill into a mountain.

Creative activities allow for this sense of incompletion to stall a project. While listening to movie directors, comedians, and writers, they all talk about how painful the editing process is. The faintest of details need to be taken care of to see their visions fully. Even when the project is released, there are details that get glossed over because deadlines need to be met. As the quote states, these great completions still come with incompletions. Do not allow the smallest of details to cloud the final project. Only the people with intimate knowledge of the project will focus on these slight imperfections. Building roads is full of small imperfections usually hidden from the public eye. Most drivers see the paved road and think of how smooth their commute is. The usefulness of the road will not be exhausted because of the minor incompletions that lie beneath.

Incompletions might include incorrect backfill on a portion of the storm sewer or improperly enclosed manholes that may leak around the pipe. These incompletions do not affect the integrity of the road.

6.4. Sales

Each sprint session is a time to sell your team on what you are doing. The picture should become clearer as time goes on, allowing your team to see your vision. Use this time to keep your team motivated. When projects or activities keep moving along, complacency can enter the equation. One project moves into the next, blurring the lines. Every week or so there is an opportunity to bring your team together and refocus them. Remotivate. Readdressing the requirements is a great way to build that focus again. Let your team know the expectations and the roles they fulfill. Some projects do not go as planned and are failures from the start. Continue to address these feelings and let people vent about what they feel the issues are. A leader takes in information and makes a decision based on that information. We do not have all the answers. Allow your team to make you aware of the challenges they are facing. Knowing they said what they have to say can motivate employees to strive for success.

Performance reviews are another powerful tool to sell your team on the goals. They have the ability to set their personal goals, and you can give them the expectations from an organizational viewpoint. Some may argue that you are taking them in a poor direction. That will be the time to sell them on your vision and how they are an integral part of your plan. They still might not be sold on the idea, and if so, then it is time to look for the right fit for them.

6.5. Lean Construction Management

Budgets have never been tighter in the construction industry. Project managers are constantly looking for new ways to save while bidding projects nearly at cost. To save money, project managers require laborers and operators to perform multiple tasks, stretching their capabilities to the maximum. Not only are the crews lean, so are the equipment and materials being used. Project managers instruct their crews to look for new ways to perform a task. If they find a way, relay that technique to management so it can be applied throughout the organization. For instance, consider a working foreman. Not only is the foreman required to oversee daily operations, he or she is also expected to run equipment, shovel, and perform any other necessary tasks around the site. On larger projects, the foreman gets to drive around in a truck to inspect the work site and fill out paperwork in a field office. While running lean projects, the foreman is expected to be out working more than doing related administrative tasks. This strategy puts a large strain on the office side, but it lowers costs in the field, resulting in better margins at the end of the project.

With projects becoming more and more complex, contractors are often required to perform extensive off-site fabrication. In the aggregate industry, crushing material down to usable sizes requires heavy machinery, and equipment repairs are frequent. Winters have become busier because of the workload throughout the construction season. In the past, winters were normally used for vacations or layoffs. Now, winters are the only time when extensive repairs can be done on the equipment. Contractors require more variations of material than in the past as well. The standard road base aggregates are still the largest quantities produced, but engineers are constantly performing redesigning projects that need different materials to satisfy their specifications. The harder-to-match specifications require special procedures and equipment that have to be fabricated. Some fabrications can be made on site, such as cutting holes in the screens to allow larger material to fall through but keeping enough material on the higher deck to stay within specification. Other fabrications, such as altering the chamber of an impactor or switching out generator engines to take on more power, need to be made off site.

Instead of sprint sessions as a typical agile project would have, these are converted to maintenance sessions. Every so often, a piece of equipment needs to be brought in for repairs. This repair is a time for the project managers to get together and see if any of the operations can be improved while the equipment is not running. The advantage of doing this in the shop rather than in the field is that the pressure to produce is not as great. The shop is for analytics about improving best practices. Brainstorming for an hour or two can pay huge dividends down the road, especially if the equipment comes into the shop every six or eight weeks rather than every four to six weeks; gaining these extra weeks translates into thousands of dollars staying in the organization's pocket rather than going into more repair costs.

Automation helps a contractor run lean as well. As discussed earlier, GPS equipment helps to limit the human resources necessary to complete some projects. While production increases with the help of technology, so does the safety of the employees working on the site. A machine can be replaced; lives cannot. In construction, safety is the number-one concern. It is why you see orange barrels everywhere while driving on the highway. Most construction sites require minimum personal protective equipment to be a hard hat, safety vest and glasses, and steel-toe boots. This protective equipment is not meant to make the employees uncomfortable, but to keep them safe at all times. GPS equipment keeps a person from having to go in between heavy equipment to take grades.

It also keeps the operator safely in the cab of the machine. While production is important, the lives of the employees are paramount.

6.6. Not Your Grandparent's To-Do List

In Chapter 5, I discussed becoming a problem finder rather than a problem solver. Problem solvers tend to receive issues, write them down, check them off their to-do list, and consider themselves productive. Here, I am going to give you a to-do list that will change your mindset once again. Agile management is meant for adaptability. Michael Norton, the former president of Zig Ziglar Corporation, has developed a way to transform your ordinary to-do list. The following additions to your to-do list will help capture what you are doing and why you are doing it for better sprint sessions:

- Write down "Why?" next to each to-do item. This helps to reframe the reason for performing the task and to prioritize which tasks need to be done. Asking "Why?" keeps your list from growing nonsensically to the point of accomplishing only half the items on it.
- Attach your to-do list items to your personal or professional goals. Asking "Why?" helps with this, but narrow it down even further to your goals. Are the items listed aligned with the future you envision? If not, remove that item from the list or, as we speak of next, delegate it to someone else.
- Delegate where you can. Add the names of people who can help or even take over the project on the to-do list. Do not feel the need to conquer all of the items on your own.

If you have the habit of making a to-do list, implement these strategies to strengthen the impact of your list. Getting things done is important, but they are not as important as getting the *right* things done. You may find your list cut in half or more. Project managers are busy people. Cutting back in any way will have a major impact. You may have fallen victim to the routine of writing a to-do list out of habit rather than purpose. Reinstall that purpose to your to-do list.

6.7. How to Grow as an Agile Leader

In this section, I will discuss how to grow into an agile leader or grow as an agile leader. Your organization may be in transition to agile management. The necessary components of your organization are a high-performing, healthy culture. How do we know what that looks like? A webinar titled "Adaptive Leaders:

Assessing and Growing Your Agile Leaders" (Elatta, 2016) by Sally Elatta is going to help us determine what to do. Elatta describes the characteristics of a healthy culture as follows.

- Clarity—Who are we and how do we behave? What is important now? Why? These are the questions to ask and realize where the organization is going. If you can answer these with no hesitation, your organization's clarity and health are in good standing.
- Focus—Can we stay focused until "done"? Does your organization start projects and then have trouble finishing them? If so, the focus needs to improve. A switch to agile or continued implementation requires focus because it will be new to many project managers throughout the organization. A loss of focus will inhibit the growth of agile in your organization.
- Predictable execution—How do we deliver value? Reliability is another word for predictable execution. Can another company rely on you to get the job done within the triple constraint? Cutting down your risk to others shows a high-performing, healthy culture.

Once your organization is deemed healthy enough to implement agile management, your leaders are going to have new roles. To be successful, agile management requires a new set of skills. Continuing the old ways will only lead you back to the old ways. Implementing a new management style requires new skill sets to transform. Elatta describes the new role of agile managers:

- "Be more strategic, less tactical
- Develop and coach people
- Own organization-level process improvement
- Lead communities of practice
- Foster creativity and innovation
- Learn the business to enable agility."

The two I want to emphasize are "be more strategic, less tactical" and "learn the business to enable agility." Becoming more strategic allows for a greater future outlook, creating improved sprint sessions. A manager who remains tactical remains in the day-to-day routine. Not seeing the forest among the trees does not allow for full growth of agile management. Agility requires seeing things before they happen. Think of a running back in the National Football League. To avoid defenders, he must think proactively about their next move in order to decide his own next move. Getting stuck in the way the linemen are blocking will result in poor outcomes. The same goes for project managers who realize change is needed in how someone is performing their work. A focus on the minutia does not result in progress. Remain strategic in your outlook. Tactical are days of the past.

Learning the business is so important for project managers. It relates to the big picture on a grander scale. See the project you are working on as a way for the company to grow. Talk to senior managers about future outlooks and their opinions. Get insight from as many managers as you can. Even talking to "rival" contractors can help you understand the direction of the industry. If they are slow, and so are you, that is the industry talking and is not a reflection of your organization's performance. Building those relationships pays dividends in the long run. People are willing to share more information with a friend than with a stranger.

Simply asking the questions can provide a valuable outlook. Some people may even give you techniques on how to improve performance. I remember contractors always trying to get our screen combinations from us. We would get aggregates to pass specifications that they could not. Of course, we would not share this information, but again, asking the question does not hurt anybody. Ask questions to learn the business to implement agile management more effectively.

6.8. "We Asked Questions. We Got Answers."

This section is based on a webinar by Lawrence Cooper titled "Organizational Agility—We Asked Questions. We Got Answers" (Cooper, 2016). Of the many questions asked to organizations, I will go over a few and discuss the number-one answers given to the questions. For more information, I recommend watching the webinar. The poll was set up in two stages. First, Cooper wanted to get open-ended answers to these questions from leaders. He compiled these responses into categories. Using these categories, he asked the questions again to formulate the answers given throughout the webinar.

First question: "Of the following, which do you believe is the greatest challenge facing organizations that want to become agiler?" The number-one answer was the internal environment. As I discussed in Chapter 5, culture plays a large role in an organization's ability to change. Agile management comes as a newer form of managing projects, leading to change throughout the organization. Old beliefs die hard. Behaviors tend never to change unless catastrophe forces it. Again, recognizing the patterns and using culture for your effectiveness helps to increase the chances of agile management surviving in your organization.

Second question: "Of the following, which single statement would you choose as a value that all agile organizations would exhibit?" The number-one answer was transparency and trust over control. Agile management is a fast-moving style. It requires failing quickly, making the changes, testing, failing, and so on. If a leader wants to control this environment, the project will get bogged down. There are too many moving parts for one single person to *need* control. This approach of letting go, trusting the system in place, and being transparent about the results helps the project. Being wrong should be encouraged. Constantly test new ideas to formulate breakthroughs. Many scientific breakthroughs occur by accident. This method of testing and failing may provide insight into something never thought of before the project started. Being transparent allows your team to see what is working and what is failing. Being mysterious to maintain control is not effective in the agile setting. You are inhibiting rather than helping.

Third question: "Of the following, which one would you choose as the most important for an agile organization to focus on?" The top answer is to embrace ambiguity. Cooper talks about the VUCA (Volatility, Uncertainty, Ambiguity, Complexity) world that organizations face now. Not only are there known unknowns, there are also unknown unknowns. Nassim Nicholas Taleb (Taleb, 2007) talks about this in his book, *The Black Swan*. Dealing with the highly improbable can take your organization to new heights. Everyone can handle known risks. That is why we develop risk registers. But handling the impacts of unforeseen events is the true test of a strong organization. Embracing ambiguity uses those unknown unknowns to your advantage. When other organizations shy away, your organization sees opportunity. That is a great space for your organization.

Four question: "From the following, which statement best represents the primary trait or characteristic on an agile organization?" The number-one answer is to create great products/services. Of course, organizations want to create great products that sell themselves. The second part of the answer is the key. Creating a great service should be as important as the product creation. Customer service when issues arise, warranties if things should break, packaging the product, and so on. All of these services create a great customer experience along with enjoying the product itself. Putting customers first so they cannot help but come back when the next generation of products is launched is the goal.

Final question: "Which of the following is most applicable in describing organizational agility to colleagues or new people you meet?" The top answer is iterative and incremental. Agile management does not have a foothold as strong as other styles. People want to know what makes it so great. Why should we take on the risk of changing from what we have now over to agile? Sprint sessions allow for repeatability that other systems do not. Seeing what works and what does not and then having the ability to change without causing a big stir is positive. Failure as feedback was the second answer on the survey. It is important to see failure in a different light. Knowing what did not work helps you get closer to the solution. Failure is not a setback.

If your organization is not using agile management, I hope this survey provides some insight as to the benefits of agile management. If your organization is using agile management, the answers should sound familiar and align with the benefits you see already. Agile is becoming more accepted yet remains a mystery for some. The change from your current system to agile will cause a ruckus. People are resistant to change until they see results. Use the Chapter 5 discussion on change, and this survey, to help speed along the process within your organization.

Chapter 7 goes to the dark side, using pessimism for optimism. It sounds contradictory, but the results pay dividends. The power of negative thinking is real. I discuss ego and its inhibiting abilities. The illusion of reality plays a large role in how we manage projects. Confirmation bias exists and is dangerous. I also take project managers into the film room. Coach Bob Knight believes heavily in the power of negative thinking and has some great insight on how to give that power life. Aristotle's "golden mean" is important to discuss because it allows project managers to remain on an even keel. Being too optimistic or too pessimistic can have its issues. Remembering the golden mean allows for both attitudes. I talk about learning to say no. This comes with practice and can empower you to make time for more important tasks. And finally, I discuss investment in the loss. There is a gain in loss. Find out how in Chapter 7.

Chapter 7

Pessimistic Optimist

As Bob Knight wrote in his book, *The Power of Negative Thinking*, "Think of game strategy the way a great sculptor looks at a slab of marble. Negative material is eliminated to create a harmonious work of art" (Knight, 2013). Erase "game strategy" and put "project management" in its place. Do not complicate things. See the project as the slab of marble and start to take away all the negative space and noise of the project. Get to the heart of the issue. Peripheral activities can wait. The project will have a critical path to assist in cutting away the extraneous activities that can wait. Use that as your guide to staying on track and remaining focused.

To remain on task, an important question to ask is "Why?" A question rarely asked and even more rarely answered truthfully. Asking "Why?" will help you to control scope. Fishbone or Ishikawa diagrams are great ways to find root causes of a problem. The question "Why?" is at the heart of those root causes. Continue to press the question until an answer is provided. Do not accept excuses as answers. Deal in facts. Take responsibility. Demand accountability.

Observation is of the utmost importance when discussing a pessimistic optimist. Not just looking, but seeing. It is the classic vase-or-face picture we have all seen: If you glance quickly at the picture, you will see one or the other. If you take the time to study the image, you can see both. Be sure to use your observing ability regularly, especially the higher in the organization you climb. When I was in college, a professor showed a black-and-white video of a group of individuals passing a ball back and forth. We were told to focus on the ball and count how many times it was passed back and forth. The question at the end was, "Did you see the gorilla walk through the screen?" Of course, I had not, because I was counting diligently. On the second showing of the video, there was a person in a gorilla costume who walked right into the middle of the video, stopped, looked at the camera, and then proceeded off-camera. This realization shook me, as I had always thought I was observant. Using suggestive focus can make the plainest-of-day actions go unnoticed. As a project manager, it is your job to keep the focus of your team on the proper things and not suggest otherwise.

7.1. Stoicism

It was for the best. So Nature had no choice but to do it.

– Marcus Aurelius, *Meditations*

At first glance, this quote appears to be pessimistic: It takes a negative tone toward an event that occurred. It sounds as if someone is asking "Why?," and the response is that it is because Nature said so. However, the optimism lies within the first portion of the quote: "It was for the best." Optimism at its finest. If the outcome is considered negative, that is because your perception of it is negative. Nature never intends negative events. It is always your perception that turns them negative. This perception is at the heart of the pessimistic optimist mindset: realizing all that could go wrong, setting up contingency plans, and then waiting for Nature to make its call.

There have been countless times when a deadline was fast approaching, and on the day we were going to pour concrete, it rained. Nature does not care about your schedule or your deadlines but always has your best interest at heart. Catching up on paperwork or phone calls or emails is always a great way to use a rain day to your advantage. The mindset shift is turning something that appears negative into the best-case scenario. Change your perception of the moment. Remember that things happen for a reason. There is a classic saying: "When one door closes, another one opens." Taking that perspective into negative situations can turn attitude around quickly.

Listening is an important skill during tough times. Most of my encounters with "negative speak" happen because people just want to vent. They want to share what they see is going wrong, and you are the sounding board. Allow that to happen. Embrace the situation. The person speaking will have an attitude adjustment just by saying what he or she feels. There are not enough occasions in the workplace when someone can talk and be heard. When I first started out, that is all I wanted: a voice.

Be a contributing member of a successful team. Realize this when dealing with junior project managers or people who work for you. Assign them tasks where their voice can be heard by the group. Use their inexperience to your advantage. They want to learn and learn quickly. No one likes being behind. Fill their empty heads with valuable information on how your organization does things, and set the performance standard right away. If there is only one way of doing things, they will not deviate from that way. When mistakes are made, that is Nature's way of reminding you of their inexperience. It is a teachable moment, not a place to point out wrongdoing. Among the best advice given to me early in my career was that if I needed to make the call, do it, and we can live with the consequences. This gave me the responsibility to make a call I felt was right, without repercussions (initially). If mistakes continued to happen, my boss would put his foot down and make some changes. The key is for those initial mistakes to be made, so you learn. Those mistakes will be the best learning opportunities for a young project manager.

A stoic's approach to the day is accepting that the people you will be encountering are the worst of the worst. Setting your mind up for unpleasant interactions throughout the day will create the attitude, "It can only go up from here." Most people are inherently good. They want to do well at their jobs, say the right things, do the right things, but people have those days when nothing is going right, and they want to make sure that gets passed on to others. Set yourself up to deal with those kinds of people. It seems like an awful way to start your day, realizing that everyone is going to be a pain, and you are the one who will have to deal with them while keeping your emotions in check. You are the person who controls your stress, anxiety, reactions, and so on. No one can make you act a certain way. It is you who has the control, not others. If you unleash fury on a team member because a senior manager put you in a tough spot earlier in the day, it is you who did that, not the team member making you do it. Knowing this can change your perspective on people and situations.

Imagine that everyone is having their worst day and act accordingly. Then, if somebody greets you warmly, it is a bonus on the day, even if that person normally does do so. Do not mistake this approach as you are the one having the bad day. You will have a great day because everyone else is having their worst day. You will be the light in their darkness. It sounds mystical and spacey, but try this approach. Do not poo-poo it because it is so different. Do it because it makes you uncomfortable. Plus, it is an internal approach. This approach does not change your behaviors, but it will change your attitude and how you deal with tough situations. Immediate stress and angst will put your team on edge, impacting their performance. Use others' pessimism for your optimism, as if you are using momentum to gain an advantage. Allow others to put the work in of stress and angst while you thrive off it.

Stoics have a spiritual practice called *premeditatio malorum,* which means premeditation of evils or negative visualization. They use this practice to envision all that could go wrong to develop responses for handling the conflict. Gary Klein, a research psychologist, developed what he called "premortem" (Klein,

2004). It is along the lines of risk analysis but focused solely on the negative impacts of risk. Premortem should be performed before the project starts. Get your team thinking about the worst possible case. What will their response be? Who will they call? Do not just get them thinking about it; develop a plan for each response. Put together a book of contacts. The project charter should have these in place, but as I mentioned in Chapter 4, risk analysis can be overlooked or not taken seriously.

Negativity is often looked at as, well, negative. But why not put a positive spin on negativity? Why is it *always* bad to have a negative outlook? See the negativity as a difference between success and failure. A surprise is rarely a good thing on projects. Project managers put effort into planning, scheduling, budgeting, and so on. Often, a project does not go as planned. That may sound negative, but it prepares the project manager to build in responses to keep the project on track. Having plan A without a plan B, C, or D is cause for failure. This technique for risk analysis is put in place to protect you from surprises down the road, especially those that are negative. Positive surprises do not need to be planned for, nor should they be accounted. Those are always bonuses that should be celebrated. By thinking negatively, it will result positively.

7.2. Taoism

> *To know you don't know is best.*
> *Not to know you don't know is a flaw.*
> *Therefore, the Sage's not being flawed*
> *Stems from his recognizing a flaw as a flaw.*
> *Therefore, he is flawless.*
>
> – Lao Tzu, *Te-Tao Ching*

"Fake it 'til you make it." I have heard that term so many times over the course of my career. I was always of the philosophy that if I did not know something, I would let you know or ask questions making it apparent I did not know. Some of the most dangerous managers I have been around are the people who do not know they do not know. Being confident is great, but being confident to the point of disillusion is not. Pessimistic optimists know they are not the best at everything and need help. They hire the best talent. They put projects in the hands of the best people for the job. They know they are flawed. Therefore, they are flawless.

A project manager cannot possibly know everything about a project or how to perform every task of the project. On a construction site, the project manager is not expected to operate equipment, because union operators on the site went to school to learn that operation. They should know the ins and outs of their piece of equipment. Not having this knowledge is a flaw recognized by the project manager and taken care of by providing talented operators skilled in the operation of the equipment. The same is true of many of the operations on a site. The foremen should have intimate knowledge of the operation, and the project manager should rely on the foremen for any questions. If there are issues, the project manager should go to the foremen for answers. However, the foremen need to recognize their flaws in project planning and scheduling. It is the project manager who has intimate knowledge of these details. Both parties should rely on the other for information they do not know. It is a team effort. Recognize your flaws. Hire talented people who are skilled in areas in which you are flawed. Create an environment of idea sharing. Build a successful team.

7.3. The Adversarial Ego

Every man I meet is my superior in some way, and in that, I learn from him.

– Ralph Waldo Emerson

The only reason for not hiring talented people is ego. You want to be the person in charge, the one who makes all the decisions. Continuing to be a student continues your growth as a person, manager, friend, and so on. You cannot know everything. The physicist John Wheeler said, "As our island of knowledge grows, so does the shore of our ignorance" (Wheeler, 2016). As you advance on a project or in your career, new challenges will arise. While your ego may have got you where you are, it can also derail your progress. Life goes on. Projects go on. People move on. Your contribution is important and leads to the success of the team, but that success is not because of you, but rather because the team is in congruence with you.

I have been on teams where the lead on the project is the end all, be all. What the lead says, goes. My way or the highway. Did that attitude make me want to work for this person and put in extra hours and effort? Of course not. In the lead's eyes, my input was not necessary. The lead had done this a million times, so move aside and listen. These types of people possess an adversarial ego. The student in them is long gone. School is for students. Work is for decision makers and problem solvers. Their ego has trapped them in an invisible box in which they are satisfied with themselves. However, the team is impacted greatly by this mindset. Meetings are very top-down, the leader speaking from the mountaintop. No conflict is generated because there is only one right answer. The team environment becomes individualistic rather than coming together. Each person has assigned tasks to do, and the team comes together only to discuss progress.

As project managers, we are the leaders of the team. We are counted on to deliver results and take the blame. That does not mean it is our way or the highway. Our teams consist of talented individuals who are there for a reason. They have seen the industry from a different perspective, and we have an obligation to include them. The more insight and knowledge we can accumulate, the stronger our positioning becomes. Team members want to work for someone who listens and takes the advice of others. Again, they make our lives easier. Being the person in charge can go to our heads and expand our ego. It is important to keep your ego in check and remain a student. You will see far greater gains from a student's mind.

Ways to prevent an adversarial ego involve introspection. Take a moment to look at what people are saying and how they are acting. Constructive criticism can come in many forms. It does not always have to be said to be heard. Are people willing to come up to you with ideas? Do you find yourself talking the most in meetings? Do your team members have a voice? The best meetings I have been a part of have had team members speaking the most, the project manager facilitating, and a designated note taker jotting away. The project manager controlled the direction of the meeting but not the message. The team was coming up with ideas and ways of doing things because they were experiencing these issues daily. They were more interactive. Learn what the team is going through and make suggestions. They are outright telling you what is happening. I have seen senior managers being told about issues, yet they remain constant in their actions. They were bull-headed, and it affected the team's performance. Their leader's ego handcuffed them.

7.4. Sales

When you do not make a sale, it is almost never because of price. Let me repeat: It is almost NEVER because of price. It is a misperception that if you lower your price, you will sell more. Imagine if Cadillac or Mercedes went to market with a car priced at $25,000. What would your perception of that car be? Did they slack on the materials to produce this model? What is wrong with this model? Why is it so low? You would try to figure out why this model is so much more inexpensive than the others. When you think of those brands, you think luxury, precision, and beauty. Would you think those same things looking at a $25,000 model? Probably not. It would feel like just another car in a long line of cars. Is that what you want your customers to think when you lower your price? The word *price* can mean reputation, goal, outlook, and so on. If you want to sell your dream to senior managers, lowering your standards is not going to help. Cutting corners is not the way to go. If you are going to sell them, give them the best product, and the best product is something they will have to pay for. As Red Adair said, "If you think it's expensive to hire a professional to do the job, wait until you hire an amateur" (Adair, 2016). That should be your mindset when selling your ideas. Mercedes does not lower its price, because it knows what it is offering. Mercedes knows it is worth the cost to own one of its cars. Lowering the price might bring more customers, but are they the customers you want to attract?

Think about a team you want to work with next. Lowering standards is going to bring in amateurs. It will attract people who can do basic things. Everyone can type and email and use Microsoft products. People have been involved in your industry for years. But those requirements do not make them professionals. If those are the standards you are setting for your team, you are attracting individuals who are basic in function. Raise your standards. Expect more. Get more. Setting that high price attracts people who will sacrifice for that goal. They know it going in, and it should offer them an opportunity to achieve success.

Some people do not want to pay that price. The payoff is not worth the effort. That makes your decision even easier. When I was enrolled as an engineering student, calculus was the class to separate the men or women from the boys or girls. It was the test to see who had what it takes to become an engineer. Calculus II was my demise. I did not want it badly enough to put in that extra work required to pass the class and eventually become an engineer. Make your team comparable to calculus. See who has it and who does not. Expect the best because you are trying to be the best. Sell your team on that vision. It is the pessimistic optimist in you. Become the bar people compare themselves to such that they will have greater success.

7.5. A Cautionary Tale

Fritz Haber, a German chemist, won the Nobel Prize in Chemistry in 1918 for inventing the Haber-Bosch process. The Haber-Bosch process is a method to synthesize ammonia from nitrogen and hydrogen gases. Food production relies heavily on this method for producing nitrogen fertilizers. It changed the world of farming for the better and continues to impact the lives of many around the world. Haber is also considered the father of chemical warfare with his work weaponizing poisonous gasses during World War I. His work in creating a process to produce nitrogen fertilizers also led to the development of chemical warfare.

I refer to this as a cautionary tale because while a project you work on may have positive outcomes, it can negatively impact its surroundings. Take the mining industry. The immediate, positive impacts are gaining resources that translate into roadways, cell phones, jewelry, and so on. Companies profit and local industry may thrive. Shantytowns are created to house the workers. The mine provides income to sustain families and keep the local economy growing. At first glance, it seems like a positive scenario. Companies profit, families sustain, and economies grow. However, the long-term, negative impacts remain troublesome. The environmental impacts alone are staggering. Erosion, the formation of sinkholes, loss of biodiversity, and contamination of soil, groundwater, and surface water are a few of the environmental impacts. Not to mention the local economies once operations seize, as shown by Wallace, Idaho.

Wallace, Idaho, was founded in 1884. Wallace became the regional center for the Silver Valley mining industry. Because of newly found silver deposits, Wallace continued to grow with new access to the railroad. In 1910, The "Big Burn" destroyed at least one-third of Wallace and killed over 80 people. Add to that the Great Flood of 1913, and Wallace started to become a place of bad luck. With each new catastrophe, Wallace would rebuild based on the unending vein of silver found in the nearby mountains. Mining began to taper off after World War II, however. Wallace's size and importance shrank with the demise of the mining industry. Today, it is a popular tourist destination for people wanting to catch a glimpse of mining life. Wallace, Idaho, and its surrounding area are considered the richest silver mining district on earth, producing 1.2 billion ounces since 1884. Wallace, Idaho, is one of many examples of a town prospering via mining and then becoming a ghost town once the source runs dry. The initial project of mining the mountains for silver ended up with rich people and reliant on a finite resource. It becomes a question of whether the risk is worth the reward. Are the 40+ years of boom worth the continued years of bust? Project managers have the same experience as this one at times.

7.6. Illusion of Rationality

This concept of the illusion of rationality came to me from Peter Sims's book, *Little Bets* (Sims, 2011). It proposes the idea that instead of making errors on the way to solving problems, people should seek data to support their points. It also hovers on the point of ideas and assumptions appearing logical on paper but not having been validated in the real world. Catering to the entrepreneurial mindset requires you to think about failure as learning. Testing becomes as important as research. You can do all the preplanning and initiating you want, but until you put that plan to work, you have no idea of the outcome. Playing the percentages leads to greater chances of success, but rarely is there ever a sure thing.

The illusion of rationality comes into play during the preplanning and initiating stages of the project. You gather the data, your experience to allocate resources, and create a plan to execute. After all of that, why do many projects fail? The process seems so rational in its attempt to satisfy the requirements. No one ever plans for a project to go over budget, beyond scope, or over schedule. An illusion is created on paper. In construction, there are thousands of pages of plans. Engineers and architects put their heads together to create a working plan and bring it to the field. When it comes time to start building, calculations don't always work out as planned, and the designs on paper cannot be completed in the field. There is always something that does not get accounted for, that needs a workaround. Hopefully, these failures are minimal and do not impact the triple constraint. If so, they can be taken as lessons learned for future designs.

Once again, confirmation bias comes into play here as well. Some engineers and architects will fight and fight for their design. They will present their calculations. They will argue how they overbuilt the design never to fail. Meanwhile, in the field, it will have failed multiple times. In Chapter 6, I mentioned building mechanically stabilized earth (MSE) walls. There are multiple ways to stabilize the earth behind the wall itself. Spraying concrete with interlocking steel mesh and driving steel pillars into the ground are two of the most popular methods. On one project, spraying concrete onto the interlocking steel mesh was the route chosen. In Figure 7.1, you will see the specification for such an installment (Missouri Department of Transportation Specification for Retaining Walls).

The dashed-lines are labeled "Soil Reinforcement." Based on the engineer's calculations and soil borings, this method was overbuilt and should have never failed. However, in the field, this system failed on three separate occasions. The illusion of rationality gave the engineer confidence in the design even though, in actuality, it was going to fail. Eventually, they went with the more stable steel pillars option.

7.7. To the Film Room

Before you can inspire your players to win, you have to show them how not to lose.
— Bob Knight, *The Power of Negative Thinking* (2013)

I titled this section "To the Film Room" because, in sports, the film room is the place for teaching. The players and coaches get together and go over the previous games looking for positives and negatives to improve performance. The pessimistic optimist approach is to look for ways to improve rather than highlight successes, or showing your team how not to lose rather than how to win. I think back to the winters in construction. It is a time for reflecting on the previous year and gearing up for the upcoming season. During this time, an organization can bring together management and field personnel to work on communication

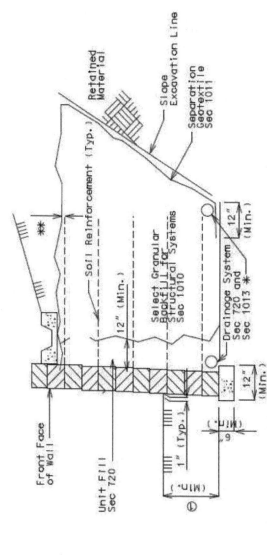

Figure 7.1   MSE wall specification for Missouri Department of Transportation (MSE Wall 04, 2005).

and ways to improve upon the processes. Inspections were always an area of concern. Each year, the aggregate producing plants would be inspected by the Mine Safety and Health Administration (MSHA). We had our own safety staff who would inspect these plants as well, yet every time MSHA landed on site, they would find violations. They do not show up to chit-chat. Their main job is to find violations, and they find them. In the year-in-review meetings during the winters, we would go over how to improve our processes to limit the number of violations. These examples were our way of showing the team how not to lose. We were much more successful in avoiding violations than accruing them, but the record can always be improved.

7.7.1. Use If-Then Statements Negatively

Legendary basketball coach Bob Knight would use this technique with all the teams he coached. Instead of saying, "If we perform well, then we succeed," he would turn it around by saying, "If we *don't* do things well, then they beat us" (Knight, 2013). This approach is an example of nailing the fundamentals. Whether it is project management techniques or the technical aspects of the work, you must know the basics to succeed. We know that remaining on schedule, on budget, and within scope are important. Losing sight of the triple constraint will get you beat. It is a fundamental aspect of a successful project; therefore, focus on ways not to lose rather than ways to win. This cognitive bias mindset shift is called the framing effect. The framing effect is how people react to a choice when it is presented differently. If you want to avoid risks, present the topic positively. A negative frame causes the audience or team to seek risks. As a pessimistic optimist, your negative framework will seek risks, then seek ways to avoid those risks. The if-then statements presented negatively will enhance the risk-seeking abilities of your teams.

7.7.2. Think for Yourself

> *As the season went along, I wanted our players—individually and*
> *as a team—to have developed an ability to think and work*
> *their own way through tough stretches on the floor.*
> – Bob Knight, *The Power of Negative Thinking* (2013)

The context of the quote is using time-outs during difficult times of basketball games. Often, when a team starts to lose momentum, the coach will call a time-out to settle the team and help motivate them. Bob Knight took the opposite approach. He wanted his teams to feel the struggle and deal with it among themselves such that they grew stronger as a team. As a project manager, how often do you receive a call that you feel the person is fully capable of answering on his or her own? The team member has not developed the ability to think on his or her own; instead, you are the safety net. A real problem develops when the project manager is not on site to see the situation and tries to make a decision over the phone. This conversation occurs when the project manager does not have faith in the people on site or has not instilled the importance to the team of making decisions on their own. At a certain point, you will have to let them drown to see if they sink or swim. Instead of calling a time-out to give them a plan, see how they handle it themselves. The result may surprise you.

7.8. Aristotle's Golden Mean

Virtue is the golden mean between two vices, the one of excess and the other of deficiency.

– Aristotle

The concept of the middle ground is common throughout philosophy. Confucius had the Doctrine of the Mean. The Buddhist philosophy conceptualized the Middle Way. Aristotle spoke of the Golden Mean. For example, Aristotle believed that courage is a virtue. Excess would bring about recklessness, while deficiency would bring about cowardice. The same is true for the extremes of pessimism and optimism. In this chapter, I have talked about ways to protect yourself. Switching questions around to find different answers means there has to be a balance or golden mean to this mindset shift. The goal is not to go from an extreme optimist to an extreme pessimist view. The idea behind this mindset shift is to find the right balance. Take the temperature of the room. Play the role of devil's advocate when optimism is running rampant. Wear the yellow hat (de Bono's Six Thinking Hats, as discussed in Chapter 3) when everyone is far too down about the situation. Balance the room.

In mathematics, the golden ratio and the Fibonacci sequence are closely related. If you do not remember, the Fibonacci sequence is as follows:

1, 1, 2, 3, 5, 8, 13, 21, 34, 55, 89, 144, 233, 377, 610, 987, . . .

It is represented visually in Figure 7.2.

Fibonacci spirals, an approximation of golden spirals, are found in nature. Examples include an arrangement of leaves on a stem, the fruitlets of a pineapple, the flowering of artichoke, an uncurling fern, and the arrangement of a pine cone. I present this information to give a natural sense to the golden mean and

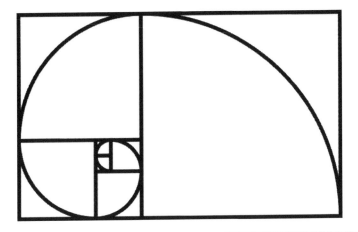

Figure 7.2   Fibonacci spiral (Fibonacci Spiral, 2016).

golden ratios. When your team and their mindsets are in balance, you can feel it. It feels natural because it is how the team should operate. As a project manager, you have to ensure that balance at all costs, continuing to find that golden mean or golden ratio. It may be applied toward resources. Which materials, in conjunction with which crews, give you the best results? More money, management, resources, and so on, may change the balance you are trying to create.

With this chapter talking about positive and negative mindsets, it is important to remember that extremes can be detrimental to your results. Constantly thinking everything will be alright can create a passive environment. On the flip side, thinking the world is going to end with every issue creates a sense of panic. You have to find the balance to achieve success. Maybe your golden ratio is slightly more pessimistic, depending on the team. Once you find that harmony and balance, it will feel natural. Many call it chemistry. It has to be developed and constantly enhanced. My grandfather often said, "Once you got it, you never lose it." Well, in this case, he could not be more wrong.

7.9. Learning to Say "No"

Pardon one offense, and you encourage the commission of many.

– Publilius Syrus

The offense, in this case, is saying "yes" because you cannot say "no." Project managers say yes for a multitude of reasons: stress, pressure, expectations, and wanting to please stakeholders are just a few. Saying yes while under pressure and stress only adds to those feelings. Seeking this approval leads to unproductive activities. Doing tasks for the sake of doing tasks does not accomplish anything. You have to know why you are doing something and what goal it helps achieve. This section will talk about saying "no" so you have the ability to say "yes."

The entrepreneurial spirit gives us the motivation to tackle mountains and scale tall buildings. Guide that motivation and angst toward positive-outcome activities. Saying "yes" to all comers may seem like a wise thing to do. You are networking with more people. You are getting your name out there. Your organization benefits as you become a go-to person for the last-minute tasks. During this acceptance, though, ask yourself, "Why?" How does this meeting or conference call or lunch out with suppliers improve your project? People pleasing comes with a cost. Telling people what they want to hear is easy. Telling them "no" so you can focus on what is important is difficult. As is the case with any other profession, practice, practice, practice. Practice the art of saying "no." Your favorite sports team does not play well if they do not practice. They need to get into a rhythm with each other to execute the plays as they were intended.

Saying "no" comes with a pessimistic viewpoint. It is a negative term. Here is the optimistic twist: Your saying "no" leads to your success. You are more focused on your goals when saying "no." The noise gets canceled. The distractions go away. You force yourself to commit to what is important. Saying "no" becomes easier when you set the rules in advance. A schedule you adhere to lets others know what you will be doing and when. If you always take lunch at noon, people will know that you will be unavailable from noon to 12:30 p.m. Again, this is your time to regroup, rethink, and recharge. Saying "no" also lets people know you are focused on only the important tasks. You do not have time to see which color works better and which font should go where. Those details do not matter. Pick one and move on to something else. Like the quote references, if you let one "yes" slip past the goalie, you can anticipate more activities pouring your way.

7.10. Investing in Loss

Give yourself to the learning process.

– Josh Waitzkin, *The Art of Learning* (2007)

As a young project manager, there is a tendency to avoid mistakes. You become cautious in your decision making, relying on your senior leaders to make the decisions ultimately. This approach to the learning process may work in the short term, but the long-term effects make you weak. As the quote states, you must give yourself to the learning process. Be all in. Learning comes from failing and observing others failing. The best moments to learn are those after a defeat. If you are winning, you must be doing something right, so what needs to change? A defeat signals something is wrong; otherwise, we would not have lost. Investing in loss means recognizing that you are going to lose, and then using that loss to your advantage. In sports, repeating a championship is the hardest thing to do. Not only do you have a target on your back throughout the season, you have also reached the mountaintop. You have accomplished the goal you set out to accomplish.

Push-back is a part of project management. Share an idea, and there are always naysayers. The natural inclination may be to defend and push back against the resistance. This resistance is to no avail. Fighting resistance with resistance only makes you tired. Becoming soft and receptive is the goal, which goes against your natural instinct. During this loss on investment, we need to retrain ourselves to handle situations differently. Come at the situation with a chance to learn instead of a chance to prove yourself. Being right is not as important as doing right. Welcoming ideas is one way to create buy-in from your team. If they feel they can come to you, you have them on your side. They want to see you succeed because, in turn, they succeed. Build a symbiotic relationship with your team.

A classic example is a new hire. The new employee comes into a new environment with new people to work with and for, and is excited. There should be an assumed loss within that employee's first few months. You are hoping to gain learning experiences, making that employee better. If you are getting the losses without the learning, it is time to look for a replacement. The new hire needs to be committed to learning, and the organization needs to back the process of learning. Lest we forget, we all started somewhere. No one came into the company running the show. Young and old, we can all learn. We must commit ourselves to the process. In the long run, the investment will pays dividends.

Chapter 8 is all about the 80/20 technique or Pareto's principle. I discuss the application within your organization, your team, your clients, and yourself. I give examples of data sets and Pareto charts to show how effective this information can be. The idea is to make every task you perform the most effective. Performing for the sake of doing something does not get the job done. The 80/20 technique focuses your efforts on what is important and what tasks create the most impact. Keeping it simple and lessening complexity is the idea behind this chapter. Performing this analysis will give you the quantifiable information to make major differences in your organization and on your projects.

Chapter 8

80/20 Technique

As Anne Lamott wrote in *Bird by Bird,* "Find the acre in which the ideas inhabit, clear out any unnecessary ideas and focus on the acre of space" (Lamott, 1995). It is about limiting distractions on your projects and within your team. Find where the success lies and focus your attention there. Find out who or what is producing most effectively and focus your resources there. Cut the fat out of your operation. The 80/20 technique will force you to find that acre of space, then find the acre within the acre, and then find the acre within the acre of the acre. Use this technique until you cannot anymore. Similar to the work break-down structure (WBS), which breaks down the project to the task level, use the 80/20 technique to the lowest denominator possible. Continue to find that acre of space where success lies, then go deep into the analysis of why it is successful and what makes it so successful. Luck and happenstance are not reasons.

8.1. Brief History

Economist Vilfredo Pareto, for whom the Pareto principle is named, was a professor of political economy at Lausanne, Switzerland, in 1893. He was interested in social and political statistics and trends. While researching the wealth and income distribution in nineteenth-century England, he discovered that roughly 20% of the people owned 80% of the wealth. Because of his fascination with the topic, he continued to test this ratio on other applications. Applying his theory to other countries and even his garden, the reoccurrence of roughly 80/20 remained constant. However, he never made the business or management connections and kept the research primarily to the academic setting (Koch, 1998). Joseph Juran popularized the principle for quality management, stating 80% of the problems were caused by 20% of the causes. He translated Pareto's principle into "the vital few and the trivial many" (Juran, Gryna, & Bingham, 1974). My application of Pareto's principle to project management will reach the client, team, and individual levels of performance. Analyze who and what are worth your time and resources as a project manager.

8.2. Application: Clients

First, apply 80/20 to your clients. See which clients are making the most money for your organization. Once these are identified, use that information to find common characteristics of the businesses. Are they large or small companies? What is their industry? What type of relationship does your organization have with them? After the first breakdown of your clientele, you will start to realize commonalities that may not have been obvious. In construction, our best clients were large companies that paid their bills on time, kept the lines of communication open, and had large accounts with us. Payment is a huge issue. If you work with a company for so long, they start to treat you as if they were your bank. You do the work to their requirements, and they think because you will do work for them in the future, you can just bank that initial work and they can pay later. Even if you will be working for them in the future, it is important they pay you for the work you have performed and not save it for a rainy day. This may seem obvious, but you will find clients for whom you have done the work, but they continue to delay their payments and become a nuisance rather than an asset.

Keeping the lines of communication open with your clients is important. For example, if there is a late payment or they cannot get to the job site as expected, you need to have a reliable contact person, someone who will get the information to the right people or has the authority to make the decisions necessary to continue work. Not only is building a relationship with the company important, you need to build a relationship with its employees. They are the true driving force behind the relationship. All it takes is one poor relationship between managers to throw away years of working together. I have seen transitional periods within companies where someone was promoted, resulting in a feud between long-term business partners. Company relationships should not be the focus. Interpersonal relationships are the backbone of any good business relationship.

It is important to focus on large accounts, because they mean long-term work, hopefully over many years. Some of the private work I have been involved with lasts for a week or two and then we would never hear from them again until the next year. A program made up of those projects creates a lot of moving parts and adds complexity that does not need to be there. Larger clients offer steadier work for longer periods of time. Being on a project for two or three months allows for the crews to get into a rhythm, find their sweet spot, get dialed in, and produce so that both businesses can make money. The ideal relationship is for both partners to make as much money as possible, not one where you are trying to get one up on each other. Working in harmony better suits both parties than a rivalry. Even if that means taking one for the team on occasion, you are in a better position, and showing your ability to compromise will better secure the relationship.

Before we begin to examine examples, the companies and dollar amounts given are fictional and used for learning purposes only. We will first examine some data and a Pareto chart to determine the companies our organization should continue to work for, and which ones we should re-evaluate. As you can see in Figure 8.1, the companies that bring in the largest revenues are not necessarily the best scenario for our organization. Belle Nation is an example. While they make a profit for us, it is not as great as it could be. They would be a company we take a second look at to evaluate how we do work with them and what processes could be made better to increase profits. The final three contractors listed need to be re-evaluated immediately, if not terminated. An organization does not have the capacity to take on losses. Find out what happened in these situations and whether they can be corrected. If not, they might not be able to be used in our future projects. And finally, look at the percentages in Figure 8.1. Does the 80/20 rule hold up? Yes: The top four contractors make up almost 100% of our organization's profits. If you stick to the general rule of 80/20, 20% of the contractors result in 70% of the profits. As a project manager, you can use that data to further relationships with those contractors, examine how the work between the two of you is performed, and look for characteristics that can help predict future results with others.

8.3. Application: Team

Applying 80/20 to your team can help you quantify who are your top performers, see what characteristics they have in common, and then use that information to find more team members like them. Figure 8.2 shows data and a Pareto chart to exemplify a way to find your top performers. See how many tasks each of your team members gets done. This number may be skewed by the importance of a task or the size of the task, but it will provide quantifiable data for your 80/20 analysis.

As you can see from Figure 8.2, the 80/20 rule does not exactly match. However, it does show your top performers. You may have already known this

	Contractors	Revenue (in Dollars)	Profits (in Dollars)	Percentage	Cumulative %
1	Pavers Co.	$30,000,000	$3,600,000	47%	47%
2	Galactic Trucking	$35,000,000	$1,750,000	23%	70%
3	Acme Co.	$20,000,000	$1,400,000	18%	88%
4	Blue Sapphire, Inc.	$8,000,000	$800,000	10%	99%
5	Road Landscapers	$2,200,000	$440,000	6%	104%
6	Red Rock, LLC.	$12,500,000	$250,000	3%	108%
7	Belle Nation	$22,000,000	$220,000	3%	111%
8	Dale's Grading & Excavating	$3,500,000	−$175,000	−2%	108%
9	Orange Barrel and Sign Co.	$9,000,000	−$270,000	−4%	105%
10	SAM Construction	$6,000,000	−$360,000	−5%	100%

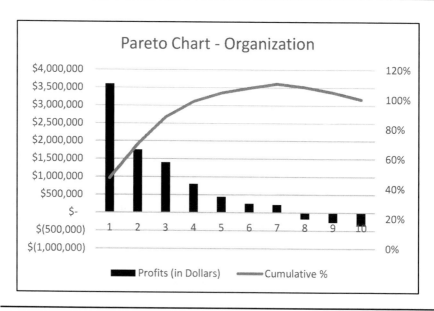

Figure 8.1   Data table and Pareto chart for the organization.

information, but now you have the data to support it. Dave, Pete, Carol, and Alyssa are your top performers. You should go to them when you are in crisis mode. They know how to get things done and account for two-thirds of the work completed. This analysis shows 40/67. Again, the minority of your team is getting the majority of your work done. Your re-evaluation should be of your bottom performers. How can you get them to perform better? In what ways can you motivate them? Are you underutilizing them?

	Team Members	Completed Tasks	Percentage	Cumulative %
1	Dave	51	22%	22%
2	Pete	42	18%	40%
3	Carol	38	16%	56%
4	Alyssa	26	11%	67%
5	John	19	8%	76%
6	Pat	15	6%	82%
7	Bob	13	6%	88%
8	Kristen	12	5%	93%
9	Belle	9	4%	97%
10	Chris	8	3%	100%

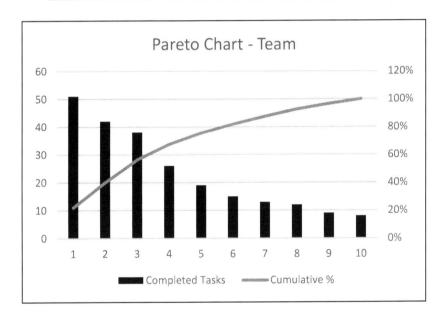

Figure 8.2   Data table and Pareto chart for the team.

8.4. Application: Yourself

Applying the 80/20 technique to your own actions can be as useful or more useful than any other application. Break down your day into the 10 or 12 most common tasks you perform and rank the impact they have on the project on a scale one of 1 to 100. Performing this analysis on yourself will help you determine which tasks to continue to perform and which tasks need to be re-evaluated. Try to be as objective in your impact rankings as you can; otherwise, this analysis will be useless.

As Figure 8.3 shows, this analysis turns out 50/80. Half of the tasks you are performing deliver 80% of your results. While eating lunch and talking to your office mates may seem important, they do not deliver enough impact on your project to warrant taking a long time to do them. Instead of going to the cafeteria or out to eat, pack a lunch so you can eat on the go. Instead of trying to

	Tasks	Impact on Project	Percentage	Cumulative %
1	Client Meetings	90	19%	19%
2	Planning for Next Day	90	19%	38%
3	Operations	85	18%	55%
4	Answer Phone Calls	80	17%	72%
5	Team Meetings	40	8%	80%
6	"Me" Time	30	6%	86%
7	Management Meetings	25	5%	92%
8	Reply to Emails	20	4%	96%
9	Lunch	10	2%	98%
10	Interoffice Relationships	10	2%	100%

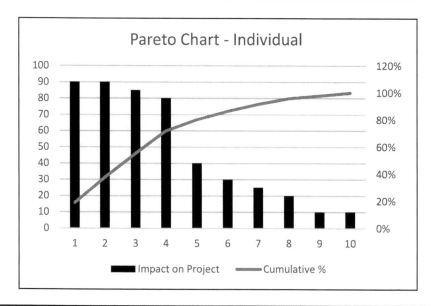

Figure 8.3   Data table and Pareto chart for the individual.

reach inbox zero, go out in the field to see how the crews are performing and if there is anything you notice changing that will bring a greater impact. Turning the mirror on yourself can bring about the greatest impact because you are the one who controls it. No one can stop you from talking about the weekend for the first two hours of a Monday, except yourself. Take a half hour for lunch instead of an hour. These changes to your day can create an immense impact on your project, team, and organization.

8.5. Cost of Complexity

The cost of complexity can be substantial. Most project managers are aware of the communication channel equation $n(n - 1)/2$. If you have five people on your team (six including yourself), the number of communication channels is 15. Double that number to ten team members, and the number of communication channels increases to 55. Doubling the number of team members increases the number of communication channels by nearly *four* times. The cost of that increase can be seen in project rework caused by communication lapses. At a large organization I worked for, their biggest issue was tracking costs. During execution, project managers were losing their ability to track time and material add-ons, or even work within the schedule, because of the project's complexity. Whether it was how much aggregate was placed that day or how many square feet of pavement was laid, quantities were not matching. If your unit price for a ton of placed aggregate is $3.00 and your number is off by 1,000 tons, you are out $3,000 because of improper scaling of the material or truckers misplacing tickets. On a $10 million project that may not seem like a lot of money, but multiply that instance over the course of the project, and it begins to add up. Executing the work becomes the emphasis, and everything else is supposed to take care of itself. In reality, tracking the project and controlling its complexity becomes a greater priority. You can place all the aggregate in the world, but if you cannot keep track of how much you placed, then you are going to lose a lot of money. Having you and your team understand the importance of documentation before, during, and after the project should be emphasized daily. Reconciling on the back end leads to you regaining portions of the revenue, but not all of it.

New software can be a valid solution to that problem. Using an in-house system works for simple projects. Microsoft Excel is used heavily for simple spreadsheets that track production and costs. In construction, most of the bidding software has an add-on that will allow you to move the quote to a tracking application. This tracking capability became a priority with that organization because there was so much work being performed but not being paid for and there was inadequate documentation to support the work performed. It is your job to provide the documentation. Never rely on another organization or client to do that work for you. Always cover yourself in the case of a dispute.

Tracking software such as Bid2Win and HeavyJob helps a project manager not only keep track of hours, equipment, and productions, it gives him or her the ability to take notes and document circumstances outside the scope of work. This information can be easily archived and accessed later if necessary. Giving your project managers this ability can increase effectiveness to levels unseen before. Instead of keeping numerous spreadsheets open and trying to update them by yourself, team members can input their production as they perform the work, and it shows up on your end. The ability to create reports from this tracking software gives senior management snapshot information on performance and enables them to look for ways to improve. The benefits far surpass data entry and historical reference. The reports generated are its true power. With one-click, you can see how your production compares to the estimated amount, how many crew hours you are working compared to the estimate, and present-value information. Having that information at hand allows the project manager to use 80/20 to determine the best team members for the task. Also, you can see what tasks your organization performs best, which may lead to a specialization not previously known.

8.6. Removing Emotions from Decision Making

An important aspect of performing a successful 80/20 analysis is taking out the emotions involved in making difficult decisions. Establish ground rules before entering the analysis. This way, if a tough decision comes up (which it will), you have the parameters on which to base the decision. Instead of telling yourself you can make the tough decision, then squabbling back and forth about whether to make it, you *have* to make the decision. Otherwise, you are going against everything that was set up before taking on this challenge. "Weak is he who permits his thoughts to control his actions; strong is he who forces his actions to control his thoughts," states Og Mandino, author of *The Greatest Salesman in the World* (Mandino, 1968). You have to stay the course. Trust in your ability to plan and act the plan. All the research is done. All the quantitative and qualitative information has been compiled. All the parameters and criteria have been established. Now it is time to take action. Remove the unnecessary aspects of your individual self, team, and organizational leaders' work. This analysis should hit every department, division, and rung of the ladder, and even task-level operations within the organization, reducing the unnecessary work at every level. Take the names out of it. Be ruthless in your pursuit. Twenty-five years of experience is not an excuse for poor productivity. Loyalty should be rewarded, but so should productivity and effectiveness. If an individual is no longer effective, it must be brought to the attention of someone who can influence the person. Knowing and not acting is worse than not knowing.

Fantasy football. To most, it is a game played on Sundays to stay more involved with professional NFL football. To some, it is a profession. Predicting who will score the most points can be a full-time job, and people have made it one. For those who do not know what fantasy football is, it is a virtual game in which you draft a team of actual NFL football players who score points for you based on their performance. A touchdown is worth six points, a field goal is worth three points, every ten yards passing or rushing is worth one point, and so on. Those point values can vary based on which league you join. You go up against another team in your league on a week-to-week basis, and the person with the most points wins the matchup. In drafting players for your team, there are many strategies. Some people want to take running backs first and often. Others want to establish their quarterback and build their team from him. A common mistake during these drafts is to build a team that is heavy on name recognition. A bunch of well-known players does not necessarily make the best team. Player A vs. Player B comparisons are made quite often with surprising results. In this comparison, names are removed to try and eliminate bias. Most of the time, it is the big-name player whose statistics do not add up to the position you are drafting him for, and you can get similar value from a lesser-known player further down in the draft. The same can go for people working in your organization. Remove the name and title. See who is creating the most value and being effective. The results may surprise you.

8.7. Stoicism

Always to define whatever it is we perceive—to trace its outline—so we can see what it really is: its substance. Stripped bare. As a whole. Unmodified.

– Marcus Aurelius, *Meditations*

Getting to the heart of the issue is the goal of 80/20. Seeing what is working, getting rid of what is not, and putting those resources toward the successes. There is no room for assuming a project is a success. You must see what it is at its core. Take all of the possible data and do an in-depth quantitative analysis of the information.

For road construction, it is very difficult to end up with partners in business. The lowest bid gets the work. The relationships established are more likely the engineering firms who provide the plans and specifications for the job. Breaking down who you work for and how successful that pairing is the goal of 80/20.

If there is a trend of failing projects with a certain firm, add more contingency to that project knowing full well you will run into issues. On the other hand, if you find successes consistently with a firm, margins can get tighter because you know they are a trustworthy source and work well with you.

When relationships are built, as when I worked in the aggregate business, the quantitative analysis can be even more revealing. We would perform half of our work for ourselves and the other half for private clients. Private clients would develop a relationship with us and use our numbers exclusively. These are the pairings you want to attack and then grow. One company we worked closely with would give us all their work for the year without even seeing our prices. They had immense trust in us to perform quality work, and we delivered year after year. Even with that being said, do an 80/20 analysis of your performance with them. Maybe they like you so much because your prices are so low compared to the market. You may be able to make more money on the deal by doing a quick analysis. Or, even with the relationship you have set up, you could be losing money on the deal. Having a consistent customer base does not guarantee you success. The price of doing business has to be right. This difference is where getting to the heart of the numbers is essential.

There is a popular show on national television on which people present their ideas to a group of investors. One of those investors cares only about the bottom line. You could bring the investor the most elaborate pitch and setup, but if you do not have the numbers to back up your claims, you might as well not have shown up to present your case. This reality is the brutal honesty it takes to implement the 80/20 analysis successfully. Numbers do not lie. If the relationship you have established is not profitable, why continue down that path? And if you do continue down that path, make sure it is because you made the changes to become profitable. Businesses do not survive on charity.

8.8. Taoism

If you take muddy water and still it, it gradually becomes clear.

— Lao Tzu, *Te-Tao Ching*

Take a step back, look at what gets results, and the picture will become clear. Three distinctions need to be made when doing a 80/20 analysis. At first, the waters will be muddy. Numbers will be everywhere if they have not been tracked properly. If you attack that chaos with fury, you will only further muddy the waters. Let the situation settle. Gather the information. Organize the data. Perform the analysis. In this way, the picture will be clearer and result in quicker findings. Abraham Lincoln said, "Give me six hours to chop down a tree, and I will spend the first four sharpening the ax" (Lincoln, 2016). The same applies for the 80/20 analysis. Take the time to collect the data and organize it in a fashion that is easily readable. Compare apples to apples. Get everything into a common unit of measure, whether it is dollars, production units, hours worked, and so on. Allow the situation to breathe so you can use brainstorming of ideas to analyze the data best. Do you want to do both private- and public-sector work? Do you want to analyze the entire organization? Find a starting point, then begin the process.

When asked to tackle a task as extensive as company-wide 80/20 analysis, some people will not be able to see the trees in the forest. They will attack it with such ferocity and end up further behind than when they started. Apply the Taoist philosophy of letting the muddy waters still and become clear. Reach out for help if the units are not comparable. Most 80/20 analyses should be comparing dollars, whether it is revenue, profit, or loss. See which area of the business is generating the most of each category and go about remedying the situation. Crystal-clear data and results need to be obtained, because this analysis will likely eliminate areas of the organization. If there is a division that has not made money in the last three years, why does that division exist? Does it provide a necessary input for another division to succeed? If so, how can we improve the sluggish division? Start to 80/20 analyze the division itself, and see where the operation is failing.

This technique can be applied to any level of the division or organization, from the top, at the portfolio level, to the bottom, at the task level. The goal is to clarify the waters not only on the technique side of things but at the organizational level. The decisions should be made clear by presenting the numbers to senior management. Initiate a plan to tackle those issues. Even in your day-to-day activities, take a look at what can be improved. What tasks get the most done? What tasks get the least done? Can you eliminate or delegate those failing tasks? If so, act immediately. The successful tasks should be amplified once the time-wasting tasks have been eliminated. Answering emails can be the biggest time waster for many individuals. Inbox zero should not be a focus. If it is, see if you can hire an assistant to browse for important emails while you continue to execute the effective tasks on your list. Once you recognize the difference it is making in your daily routine, ask the team to do the same. Get everyone thinking about what they are doing and if it is effective. If not, how can they make it effective, or how can they eliminate the task?

8.9. Sales

The 80/20 technique is the best for selling management on an idea. It provides the concrete, quantitative analysis management similar to the qualitative data of what qualities you look for in a team member. Showing management how you can lower costs while remaining productive makes your ability to sell much stronger. Every business strives to lower costs while increasing production. This technique revolves precisely around that idea: getting rid of the waste in production or management while maintaining or increasing the production from each area. A company I worked for had a prime example of how to muddle the idea of management. They created a division focused primarily on management, similar to a project management office. One manager from that division would oversee a project, while a different division would have a manager for their work. Each of those managers would in turn report to a different manager. The project managers would have a foreman/supervisor in the field to manage operations. On top of that, there would be a manager assigned to control human and equipment resources. For one project of small to medium size, there would be up to ten managers working on the project. That may seem normal for some industries, but it is not the case here. No one would know who they needed to talk to when an issue arose. No communications plan was established. One of the first things to resolve was the many communication channels. Subcontracts would get confused, and payments would get delayed up to six months after completion of the project.

For most, that example of adding more managers to a problem may seem rational. However, if an 80/20 analysis was performed, management would have found out who the actual people running the project were. Instead, they kept adding project managers to the problem, causing more problems. The importance of stepping back, taking a deep breath, and re-evaluating the situation cannot be overstressed. In construction, the main concern is getting the work done. Tracking and managing on the back end are often overlooked, because the schedule is the main driver of the project. Get the work done as quickly as possible and worry about the other items later. Try to reconcile quantities as you go, but it can always wait until the end. It only leads to trouble. Make sure your project managers are doing an 80/20 of their time management. See how they are spending their time on which activities. Answering phone calls and emails is great if they are answering pertinent questions. Otherwise, it is wasting time they could be applying to tracking quantities and getting paid. Stress the importance of this approach in performance evaluations. Do not let it slip. Show them your analysis of yourself to put words into actions. This analysis is your opportunity to sell them the importance of this analysis and how it will pay dividends. Redlining yourself only leads to burnout. There have to be periods of idle time that allow reflection.

Whenever there is a meeting for senior management to discuss projections or future outlooks, "next level" is a term I constantly hear. They want the group to take their performance to the next level. I have yet to find out what that means. Rarely is "next level" defined in the meeting. It is a buzzword that senior management loves to use to try and inspire the group. My goal for my team is 10×, which means to multiply the initial goals by 10 and have the team shoot for the moon. It has been popularized by Grant Cardone, best-selling author and real estate mogul (Cardone, 2011). By using this 10x approach, people force themselves to dream big, bigger than they had ever imagined. It is a nice, round number that is easily multiplied to your goals. Add a zero. We can all do that. It forces the individual to take on a goal that is nearly impossible but will take that person to the "next level." Quantifying goals is important for your team as well. In addition to using 80/20 analysis to quantify, you are also taking those results and trying to 10x them right away. Once you find what you do not need to do, it may be your 10x goal right there. Then your "next level" was too easy. You have to continue to 10x and 80/20 after each milestone or stage gate. This constant analysis along with far-reaching goals will ensure your team continues to progress successfully.

Think about a time your goal was too easy. Was it satisfying to hit that mark? Or did it come and go with little fanfare? Probably the latter, because you knew you could hit it. Now take that same goal, 10x it, and see how you feel after reaching that goal. Elation. Surprise. Pride. Some of the words to describe your feelings. Even if you fall short of 10x but land on 3x, that is next level. Shooting for the stars and landing on the moon is better than jumping to touch the door frame. One is easily managed; the other is nearly impossible. One goal keeps you grounded, the other puts you in space. Which goal do you select? Do not be one of those managers who settle for their teams producing just enough to satisfy the requirements. Satisfy those requirements in record time. Make the customers so happy they do not even think about another contractor the next time they need a project done. Some teams may not be ready for this level of advancement. They must reach the performing level of team building and then fuel that fire. For other teams, the "next level" may be norming. If your team cannot get out of storming, do not start talking 10x and reaching for the stars. Get each of them on the same page first. Walk before you sprint.

8.10. KISS Principle: Keep It Simple, Stupid

"If you make most of your money out of a small part of your activity, you should turn your company upside down and concentrate your efforts on multiplying this small part," wrote Richard Koch, the author of *The 80/20 Principle* (Koch, 1998). This idea seems like a drastic move. Analyzing your company, finding that a small part of it makes the most money, then flipping the company entirely on its head to multiply this opportunity. But does this approach make sense? Why would you continue to forge ahead with ventures that are bringing you down? You can apply this idea not only to a company, but also to a team, an individual, or yourself. Segregate the best attributes, focus strictly on them, and start to multiply them over time to attain the best possible result. What is the definition of insanity? Performing the same task over and over and expecting different results. If you continue to dabble in the retail space when wholesale is where you are making money, why would you continue to stay in the retail space? Find what works and do it better than anyone else.

The KISS principle enters by simplifying your structure to what works and working smarter, not harder. In construction, there is constant machismo. Lifting a manhole grate by yourself does nothing more than demonstrate your strength. It does not benefit the team or the task at hand. Technically, you are working harder but not smarter. Use a skid steer or have someone help you, so the task takes less toll on you. The smarter approach may take a minute longer. Use that minute to plan the next move. On every crew I worked with, work smarter not harder was a mantra I repeated. There are constant opportunities to dive in and start lifting or breaking or hammering. Take the time to approach the situation smarter. Use the finite energy you have to apply for greater gains rather than ego expansion.

Businesses are using the KISS principle as well. How many advertisements have you seen in which people are selling mattresses, clothes, shoes, and so on, online only? No more brick-and-mortar stores. No more salespeople. Instead, the approach is to cut out the middle and go directly from the manufacturer to the customer. This approach is simplifying a business model. Instead of trying out the mattress in the store while someone stands over you waiting for a reaction, you can now try it out at home for x number of days. If you do not like it, you can return it. Clothes and shoes are going to a subscription-based model where you give them your style and measurements, and they send you outfits to match your preferences. This membership simplifies your shopping experience and simplifies the amount of overhead the clothing store carries. All of these ideas are taking old business models and simplifying them to the essentials.

Size and complexity are so intriguing. For instance, assume you are at a dinner party. As you are mingling, people ask you what you do for a living. Which answer sounds more impressive? "I am a senior vice-president of a software company overseeing the research and development of cutting-edge technologies," or "I run a small brick-and-mortar retail store." The senior vice-president sounds amazing. The job probably comes with perks we did not even know existed. The brick-and-mortar store sounds mundane. Size and complexity are all about ego. We want to run the biggest operation available, whether it involves supply chains, distributors, contractors, software, automation, and so on. If we can handle all of that, we must be very smart. All the while, our operation may be in the red two years running, while the brick-and-mortar store is in the black every year because of its simplicity. It sounds mundane; the curb appeal may be low, but it outperforms these conglomerates. Why? Because of simplicity. Do not let ego affect your decision making. If it is not working, scrap it. It is better to lose early than late.

Richard Koch comments on complexity in his book, *The 80/20 Principle.* His observations are as follows:

- "Waste thrives on complexity; effectiveness requires simplicity.
- The mass of activity will always be pointless, poorly conceived, badly directed, wastefully executed, and largely beside the point to customers.
- A small portion of activity will always be terrifically effective and valued by customers; it is probably not what you think it is; it is opaque and buried within a basket of less effective activity.
- All organizations are a mix of productive and unproductive forces: people, relationships, and assets.
- Poor performance is always endemic, hiding behind and succored by a smaller amount of excellent performance.
- Major improvements are always possible, by doing things differently and by doing less." (Koch, 1998, p. 95)

The goal of these observations is to eliminate low- or negative-value activities. Reduce your plan of attack to the essentials. Size and complexity, in and of itself, do not add value.

8.11. Capacity

After running 80/20 for your organization, team, and self, changes will be made. Sandrine Provoost, the founder and CEO of Vanguard Change Consulting, presented a webinar titled "Preventing the 10 Common Mistakes in Leading Your Change and Transformation Projects" (Provoost, 2016). One of the mistakes she pointed out is capacity. Provoost explains capacity as "not creating adequate capacity for the change, setting unrealistic, crisis-producing timelines, and then laying the change on top of people's already excessive workloads." This quote should sound familiar. You are busy trying to run projects, and now you have to transition while performing your day job. She recommends performing a *change capacity* review. Any capacity required for change is not available for operations. This issue will place emphasis on the change you are trying to make. Instead of finding ways to sneak a few minutes here and there, you need to create blocks of time dedicated to change. As with any change, personal or professional, if you do not dedicate yourself to it, the change will not occur.

So how can we free up time during our busy day to implement change? Provoost has developed five ways to free up capacity required to make the change:

1. Take work off/stop work, slow work down; read just timelines.
2. Pause work; put it on the back burner.
3. Reallocate existing people with best skills to priority efforts.
4. Hire the right skills and knowledge.
5. Outsource work, use external experts to do appropriate work.

The technique you use will depend on the situation. This decision is where 80/20 analysis will play a large role in deciding which activities can stop or pause, who is your best performer, what skills work best for each position, and even the outsourcing of work. If you use 80/20 analysis for each technique, you will have time available to make the desired changes. You may find your priority efforts changing after the analysis. As we have seen, 80/20 analysis can be performed at any level of your organization as well as any level of the project. It has a multitude of applications that should be used frequently to maintain a successful path. I hear people complain about how busy they are and now they have even more to do. Using these five suggestions along with 80/20 analysis should highlight where your priorities lie and what needs to be done next.

Chapter 9 discusses internalizing outcomes, both successes, and failures. As a project manager, you will prosper and you will flounder. Internalizing these outcomes has negative consequences. Your ego becomes inflated in victory, and your heart breaks in defeat. The ebbs and flows take a toll. The idea behind this chapter is to make a decision based on information, and it is what it is. I talk about the difference between being lucky and being good. Is there a difference at all? I discuss putting your trust in the system. You have done the planning and research, now trust the information. Celebrating the wins is important. The tendency is to brush wins under the rug and focus on the losses. In doing so, the negativity builds without any positivity to offset it. Realizing that both parties can be correct helps to keep the argument external. This chapter is about self-awareness. Make a decision and let the chips fall as they may.

Chapter 9

Internalize Failure and Success

One of the worst mistakes I made on a quote involved a recycle concrete pile. A nearby road was being removed and then replaced. The removed pavement was being stockpiled off-site, and it was our job to crush that material into usable material for the road project. After speaking with senior management about the quote, I had two quotes in hand, one if it was good, and one if it was bad. However, I did not have the quote for ugly. I turned in the bad quote, and we ended up with the project. After opening up the pile of recycle concrete, it was nothing but garbage, dirt, plastic, and so on. Hence, why I should have had the ugly quote along with me. We ended up compromising on a price with the owner of the project. It was an oversight at every level, but ultimately I was the one with my name on the quote. If I internalized that failure, I would have lower confidence in putting together quotes for future clients. The mindset has to be the same as when a pitcher blows a save and loses the game for his team: Get back on the mound the next day and bring the team to victory. If you sulk and let the failure consume you, you will no longer be as effective to the success of your team as you need to be. They depend on you, in good times and bad, to bring your best each and every time. If you get knocked down, get up. Some failures are harder than others, but everyone experiences them. Take it as a lesson learned and never repeat the mistake again.

The failure mentioned above was a team failure, even though I was accountable. It is a change of mindset that needs to happen. It went through many levels of inputting, scouring, and reviewing before I handed it in. It is similar to a boxer losing a fight. All the audience sees is the one person in the ring taking the loss. No one sees the team of individuals who helped prepare that person for this one opportunity. Should the boxer lose, it is a team loss. The game plan they established did not work. The workouts performed in camp did not prepare the fighter for the task at hand. When a project manager sends a team member on a task, make sure the team member can perform that task. Do you have all the information necessary to make a proper decision? Have you trained the team member for this task? If not, what can you do to make your team more prepared? All of these things go into an individual's failure. If a failure does occur, take that person aside and break down the thought process and present solutions. Do not overreact. Do not make a public scene with the team member. Bring the team member into your office for discussion. Get feedback. It is important not only for the individual but also for future team members. It will make you aware of your shortcomings when training people. It will make you stronger as a project manager, and it will make the team stronger.

Over the years, I have noticed people personalizing projects: "I built that bridge." "I built that road." It is never "My team built that bridge." "We built this road." Exclusive terms like "I" mean an internalization of the success as occurred. No *one* person built anything as large as a road or a bridge. A team of hundreds built those projects. A project manager may have *led* the project team but did not *build* the project. Keeping a team attitude is important. You can have responsibility and accountability without internalizing it. RACI (Responsible, Accountable, Consulted, and Informed) charts are effective to keep people from internalizing their roles. If you are held accountable, it is your job.

Loss does not equal wrong; win does not equal right. I have seen so many people consider themselves good at their job because they happen to be right. Ever been out with a friend to watch a game, who predicts the final score correctly and then thinks he is the best at predicting scores? He gets it right one time, and it validates him. Be careful of this pitfall. It is classified as a subjective win. There is no rhyme or reason behind the correctness, it just happens to be right. Basing decisions on a subjective win or loss can be devastating. It creates a personal attachment to both positive and negative outcomes. The rollercoaster of emotions begins to happen. You live and die with each result. You may even experience the five stages of loss (denial, anger, bargaining, depression, and acceptance).

An objective win or loss is where you want to be. Do the research. Compile the results. Make an educated decision based on the information. The win or loss will be based on facts, helping to reduce any personal attachment to it. This keeps an individual from internalizing the results, good or bad. You are the decision maker, but you are not the reason for the result. If you possess a crystal ball and the ability to predict the future, then the result will be because of you. You knew the outcome. I have yet to meet someone with that ability. You can predict with a high level of certainty for a particular outcome, but you can never *know* the outcome. Keeping this in mind will help to reduce the impact of negative results on your future decision making.

9.1. Stoicism

Things gravitate toward what they were intended for. What things gravitate toward is their goal. A thing's goal is what benefits it—its good.

– Marcus Aurelius, *Meditations*

If you continue to put in the hard work and make the sacrifices to become the best at your craft, the work will pay off for you. A goal will help you recognize the effort you are putting in and become a realization. It sounds hypothetical, but think of the goals you have achieved in your career. They did not appear out of thin air. You worked to achieve them. Success wants to be achieved. Help it become a reality. Rarely does a person on a project want the project to fail. It fails because the work required for success was not done. This mindset helps to keep you working toward your goals. Realize that success is possible by putting in the work. Do not get fooled that because you are on the project, it will succeed. Past performance does not predict future results. This statement has been uttered many times, yet people continue to internalize their successes, feeling they are the reason for what is happening. The reason for success depends on many variables, including your actions, but your actions are not the only reason success happens. Your team and how they perform will influence a project far more than your own actions. You could develop the greatest plan and budget in the world, but if your team is not talented enough to perform the work, your mastery goes to waste.

Have you ever been to a conference and gravitated toward an individual for some reason or another? That person might be funny, charismatic, good looking, and so on. Whatever the reason, you find yourself wanting to be around that person or that group of people. The same goes for success. It gravitates toward certain individuals. Why is that? Look back at the quote at the beginning of this section. Success gravitates toward its intention. Successful people build systems that ensure success. It is not a guessing game. Successful people know how to be successful and how to create success. One way for them not to internalize results is the failures they have experienced along the way. Without the downs, there are not any ups. It is comparable to a rollercoaster: You have to go down to get back up. Internalizing a success would keep them down for every failure they encountered, blaming themselves for the failure instead of the system they created.

9.2. Taoism

> *One who boasts is not established;*
> *One who shows himself off does not become prominent;*
> *One who puts himself on display does not brightly shine;*
> *One who brags about himself gets no credit;*
> *One who praises himself does not long endure.*
>
> – Lao Tzu, *Te-Tao Ching*

Each line of the Taoist philosophy above represents an individual internalizing a result. Take the first line about boasting and not knowing what you are doing. Earlier, I mentioned the phrase "Fake it 'til you make it." I am not a big believer in this, and the quote above helps to cement that belief. The loudest mouth in the room is trying to prove something. Trying to talk the talk rather than walk the walk, which transitions right into the second line of the Taoist Way. Being loud will get you recognized, but for all the wrong reasons.

I also do not believe that any publicity is good publicity. If I get recognition, I want it to be for a success I have realized or my ability to accomplish something great. Epic failures including bridge collapses, squandering millions of dollars, or project overruns by years gain recognition, but it is not the kind of recognition I want to be associated with in any way. Keep a humble mind in both victory and defeat. Internalizing both will keep you on a crazy rollercoaster of emotions. This also applies to being a good teammate. Do you want to help people who constantly talk about how great they are? They are probably the last people you will want to help. If they are so great, why are they not doing it themselves? That mindset is toxic to a team but comes easier if a person in the group feels he or she is above the group.

The last line in the quote exemplifies the internalizing of failure or success. If you are continuing to tell others how great you are and how special you have become, you will not long stay with the company. Being talented and successful will garner attention on its own. Self-promotion is not a bad thing, but be careful with the power it holds. It can help individuals reach status unthinkable, but with that comes a great fall if arrogance prevails. In this book, I have talked about being the "white rabbit" of your organization. Be recognized for your talents and stand out from the group. This does not mean you should toot your own horn at every conceivable opportunity. It means to work hard toward your goals and have your work speak volumes. You will become recognized for your talents of performing rather than your talents for talking. Who do you want on your team? The talker or the walker?

9.3. Sales

Focus on the people, not the project. How many times are you so caught up in the details of the project that you forget the needs of the owner? It happens more than it should. Keep the focus on the requirements and needs of the owner. Having a deep knowledge of the project and all it includes is important, but more important is pleasing the owner and meeting his or her expectations. The needs of the project must meet the needs of the people involved. This perspective is important when selling an idea. Make it about others rather than yourself. Do not internalize the project and make it yours. You are the facilitator making the moves necessary to perform the work. Nothing more. Nothing less. Create the focus among your team. Make them realize that if the owner is happy, they should be happy. You can do more preparation than anyone else, but if your work does not apply to the problem and solve it, no one will listen, and no one will want to buy into your idea.

Think back to the last time you bought a car. Did the salesperson ramble off all the specifications of the engine? Were you interested in the off-road package? If not, did you learn more about it than you ever thought possible? These are the things you must keep in mind when trying to sell your idea. It may interest you that the car goes from zero to 60 mph in three seconds, but if all the client wants is something that gets good gas mileage and is dependable, then speed is not an issue. Be the best listener you can be. Hear what the client wants and play to the client's interests, not yours. You may think it is the best car on the market because of its performance, but the buyer normally commutes ten minutes to work and stays local on the weekends. Or the opposite may be true. The client may not want the more economical version of the car. The client may want something that goes from 0 to 60 in under three seconds even if it gets only 10 mpg. Think of the reaction if you try to sell that person an economical car that gets great gas mileage on the highway but takes a while to get there. You are going to lose a client because you are not listening to the client's needs.

The same is true in project management. Meet the requirements of the owners. If they do not want a refrigerator in the break room, do not give them a refrigerator in the break room. If they want stained-glass windows instead of regular windows, you are going to put in stained-glass windows. Technology can be a particularly difficult case. I remember when my father was attempting to purchase his first computer. He watched all the infomercials and read all the articles describing the newest processor speeds and bigger monitors. He did not want to make a purchase and then have a newer model come out days later. So he waited and waited, until finally he decided to buy one. He had realized that after he bought one, he could upgrade as time went on, and that would satisfy his need for the latest and greatest. He needed to hear that getting a computer was the easy step, because his major concern was the upgrades. Again, some people get caught up in the details without recognizing the easy solution. And if you struggle to make those connections, take copious notes on each of your interactions. See how the client reacts to what you are saying. This note taking will help you get a feel for what your clients' needs are and how to respond to them. Do not internalize your successes or failures. Learn from both. If you succeeded, find out why. What is it you said that sealed the deal? If you failed, why did you fail? Never leave a result unanalyzed, good or bad. This analysis will allow you to build a system of success rather than grasping at straws.

9.4. Are You Lucky or Good?

What is the difference between a lucky person and an unlucky person? This question is not a joke. Truly think about the difference. Why does one person catch a break and the other falter? Is there a large discrepancy between the two? Both work hard. Both have a goal. Both have similar resources. But one gets the job, and the other does not. This example is why you should not internalize success or failure. You are not much different than the person who keeps failing. And if you think you are, you are bound to fail . . . impressively. Many factors contribute to one's success. It is less about you and more about your circumstances. I consider myself a lucky individual. I had a car at 16. I graduated college without student loans. I have worked for successful companies and consider myself a success. I had much the same opportunity as anyone going to my high school. Laboring for a construction company is not a difficult job to land, but it is a difficult job to stick with and make work for you. That is how I afforded college. Most, if not all, people can successfully apply for and accept a laboring position. I worked hard at that job, but that is not why I graduated without loans. I saved the money I earned and applied it directly toward those loans. There are many aspects of my life where I think I am lucky but my characteristics are not otherworldly and can be done by most people. This example is the mindset you need to have when success comes your way. Of course, you put in the work and learned the tools to invite success to the party, but that is not why you achieved success. Remaining humble will lead to more success and benefit more people than if you think your success is because of you.

9.5. Trust the System

In his book, *What I Learned Losing a Million Dollars,* Jim Paul started to internalize everything, good or bad, that fit his narrative (Paul and Moynihan, 2013). If a sale went well and he made thousands of dollars, it was because of him and his expertise. When a bad day came, it was because the markets performed poorly and not because of a move he made. Whatever the result, he was internalizing it to fit what made him feel best about his abilities. He discovered the best way to deal with this was to remain objective and remove as much emotion as possible. Excluding feelings and emotions allowed him to make decisions that best suited his situation. If he was down, he was not going to make a rash decision to try and win back the losses plus gain on the day. "There is not a six-point shot in basketball" is popular because teams will get down and start to press. They will force the issue. They are trying to score more points per possession than are possible. The idea is to slowly cut into the lead and not win the game all on one possession. Keep running your offense and getting stops on defense. Paul realized this well after losing a million dollars. Stay the course. Trust the systems you put in place. They are there for a reason: to keep you from making irrational decisions trying to win it all in one day. Projects are marathons with nuances.

The project charter remains your Holy Grail for information during projects. All of the preplanning and requirements are found in this document. Refer to it often. The systems put in place will keep you on track. Project managers rely far too often on their memory as opposed to maintaining documentation and having a quick way to reference it. I have seen people go into meetings without anything to write on, as if they are going to remember everything that is said. I can barely remember what I had for lunch yesterday. How do these people expect to remember important takeaways from a meeting? This difficulty is why putting systems in place for reference is so important. Whether it is expectations for the team's behavior or requirements from the owner, the project management plan should have these documents.

9.6. Grow and Achieve

"Elimination of ideas that have outlived their utility is essential to almost any process of growth and achievement," wrote Bob Knight in *The Power of Negative Thinking* (Knight, 2013). This statement can relate to an organization's best practices, an individual's idea of how things should be done, or a team member's suggestions. As I have mentioned previously, "best practice" is a term an organization uses to stunt its growth. It gives the impression that no further work needs to be done in how we do things. This technique is the best way to do it, and that is all that is required. Companies can use the same best practice for 25 years and think it is still relevant. It is possible that a 25-year-old technique may still have relevance, but my guess is that it has far outlived its utility.

The same goes for watching people work. How often have you seen team members try to perform a task, and they are not sure how to proceed and may even seem clueless? Instead of watching them struggle, give them ideas on how to perform better. In construction, I see it all the time. A laborer with very little experience will try to use a tool, and the crew will stand around to poke fun. They find the struggle amusing. Why not teach that person and help him or her learn, as opposed to finding humor in the difficulty? It is a mindset that can be set by the project manager. If they see you are going over there to help and teach, eventually they will start to do the same. It starts at the top. The stand-around-and-watch-someone-struggle mindset needs to be eliminated. Then watch how quickly the team will grow and achieve.

"Addition by subtraction." We have all heard it. It is a contradiction that makes sense. By eliminating the bad apple, team morale and performance increase. I had a foreman who was telling everyone how poorly managed our division was becoming and how it had changed since the new powers-that-be were put in charge. Every day, he would come to work with an attitude that would poison his team and those around him. Whenever someone would ask him how his day was going, his answer would always devolve into a discussion about how unhappy he was with the new regime. Criticism can be constructive and often necessary. It helps to build up the weaknesses. In this case, it was the same old stories. The continuous attitude led to him being let go because of the toxic atmosphere he created for those around him. Once he left and a new foreman was in charge, production increased and the employees were visibly happier. It was not a coincidence. The old foreman's utility had run its course, and to grow and achieve, elimination was the only answer.

9.7. Celebrate the Wins

It is easy to say, "Do not keep score." This chapter has talked about removing the ego and self from the outcomes. This section is about the transition period from constantly keeping score to fully removing one's self. As success starts to come, celebrate those victories. When was the last time you celebrated? When was the last time you sulked after defeat or failure? I would guess that the answer to the first question is longer ago than the answer to the second question. We beat ourselves down without picking ourselves back up. If we are going to keep an internal score of failures, then we need to celebrate the wins. Eventually, the score will not matter.

As more successes come, the victories will seem a foregone conclusion. Expectations will be placed on you to reach those levels again. This situation is where scoring victories at zero and defeats at full credit will cost you dearly. Scoring in that way will always result in a loss, not only for one result but for every result after that. Bill Walsh was one of the most successful coaches in National Football League history. He won three Super Bowls with the San Francisco '49ers, taking his team from 2–14 in 1979 to their first Super Bowl victory in 1981. After he won his first Super Bowl, expectations for his team were always high. Making the playoffs was not enough. Winning Super Bowls was the only acceptable outcome. For Walsh, in his book, *The Score Takes Care of Itself* (Walsh, Jamison, and Walsh, 2009), his ways of dealing with escalating expectations and personalization of results were these:

- Do not isolate yourself: Have a small group of people going through similar situations, a crying shoulder to understand the pressures.
- Delegate abundantly: You have hired talented people, now put their talents to work for you.
- Avoid the destructive temptation to define yourself as a person by the won–lost record, the "score," however you define it.
- Shake it off.

As a project manager, you are not on an island to solve and find problems on your own. Use your resources. Ask for help. Believe it or not, other people have been through similar, if not the same, situation you are going through in your work. If work continues to pile up, reach out—asking for assistance is not a sign of weakness. It shows great strength and awareness. Inherently, people want to help. Finally, develop a tough skin and shake it off. "To accept it without arrogance, to let it go with indifference," wrote Marcus Aurelius in *Meditations*. In a way, what he is saying is similar to "water off a duck's back."

9.8. Can Both Parties Be Right?

When project managers begin dealing with management and other stakeholders, there will be disagreements. These disagreements are about requirements, scope, schedule, budget, and so on. Is it possible that the project manager and the stakeholder can be both be right even when they disagree? Here is an example from a medical textbook, *Epidemiology*, by Leon Gordis:

A physician visited his primary-care physician, an internist, for an annual medical examination including a stool examination for occult blood. Of the three stool specimens, one tested positive. Based on the internist's experience with false positives, he said the positive test was of no significance. The internist ran three additional stool samples, all of which came back negative. The physician-patient was still skeptical, so he went to a gastroenterologist for further examination. Based on the gastroenterologist's experience, any positive stool finding is serious and is associated with a pathologic gastrointestinal disorder. Gordis goes on to explain that both the internist and the gastroenterologist are correct. In general medicine, populations have a lower prevalence of any serious gastrointestinal disease, while the gastroenterologist sees a population with much greater risk (Gordis, 2009, p. 103).

Being correct may be a matter of subjective experience. A general contractor will have a much different perspective on an issue than the contractor who deals with this issue daily. Some people even have the ability just to look at something and recognize future issues. Houses come to mind. Renovators may look at a house and know it has lead pipes because they have seen this style of house many times, and it is a common occurrence. In the medical example, the internist has such a busy clinic and sees so many false-positive results that another string of tests coming back negative is all the internist needs to say it is not a major problem. However, the gastroenterologist sees a specific population in which a positive test is very often serious, because the gastroenterologist is a specialist who sees mostly serious cases.

Factoring in your experiences with the experiences of stakeholders can result in disagreements in which you both believe you are correct. Based on your experience, this is where you might start to internalize that this stakeholder is difficult and wants to argue about everything. You have to realize that the stakeholder is correct, and so are you. Disagreements do not have to be black and white, correct and wrong, or this or that. They can be both. Disagreements are often nuanced with plenty of gray. If you are one to have to win, often you lose. Your ego becomes your identity. Your team will pick up on it, and then you are someone they consider difficult to work for and someone to avoid. Agreeing to disagree can help save you time and effort that can be used elsewhere to make a real impact on the project.

9.9. Narcissistic Injury

> *In life, there will be times when we do everything right, perhaps*
> *even perfectly. Yet the results will somehow be negative. . . .*
>
> – Ryan Holiday, *Ego Is the Enemy* (2016)

A narcissistic person is defined as having an excessive or erotic interest in oneself and one's physical appearance. Self-centered, egotistical, and conceited also describe this individual. A narcissistic injury refers to any threat to this person's self-esteem. This event can be inconsequential to most people, but for the narcissist, it becomes a personal attack. Being right is a quality of a narcissist. Any challenge to that person's opinion or recommendation turns into unnecessary conflict. The idea behind this section is to point out that being right and doing right are two different things. Doing right can lead to negativity. According to Ryan Holiday, these negative characteristics are failing, being disrespected, feeling jealous, or no reaction at all (Holiday, 2016). There are times in construction when getting things done prevails over doing things right. These occurrences do not affect structural integrity; they are more often an attack on aesthetics. Doing right is an external action and is a selfless act for the betterment of the program, project, team, and so on. Being right is an internal action and is a selfish act for the betterment of your ego.

In this chapter, the goal is to make decisions externally and keep the outcomes external as well. Doing right helps keep the outcomes external. If you know that what you did was right for the betterment of the project, you cannot beat yourself up over the decision. As the quote at the beginning of this section states, you can do everything perfectly and still not get the desired outcome. I will relate this to football. How many times have you watched a wide receiver break free from the defense, only to have the quarterback overthrow him for an incompletion? Execution on the route was perfect. The play still failed. If you are the play caller, you did your job. You did right for the team. The outcome was still negative. The play caller should focus on the work itself instead of only on the outcome. This a personal mindset change. The organization or team may think differently, because the focus remains on the outcome. If you were to internalize strictly outcomes, there would be times things just happened to work. You may have forgotten to schedule a subcontractor, but it rained that day, so you got lucky. Does that make you a better project manager than if you did schedule the subcontractor but they put you behind schedule? Forgetting resulted in a better outcome than scheduling. A narcissist would feel better about the outcome than about doing right. Doing right and putting the project behind schedule would result in a narcissistic injury. Change the focus from the outcomes to the process. Do right instead of being right.

9.10. Biases: What They Are and How to Use Them

This section is going to point out biases people have when making decisions. These biases are internal and even subconscious. Bringing about the awareness that these exist can help a project manager make a better decision. Identify, analyze, and act. Those three steps will help improve the decision-making process for you and your team.

9.10.1. Anchoring Bias

An anchoring bias describes the attachment some have to the first number they hear, see, or read. In construction, each project will have an engineer's estimate of how much the project should cost. All contractors base their estimate on this number. They are "anchoring" your decision making on the engineer's estimate. What happens when you go through the project and realize that your number is 10% higher than the engineer's estimate? Do you change the number to fit the anchor, or do you stick to your pricing? Some contractors become so fixated on the engineer's estimate that they try to get closer to that number instead of pricing the project based on their own information. Do they assume the engineer cannot be wrong? This question rarely gets answered, yet the engineer's estimate becomes gospel. The way it should be looked at is a ballpark number, not an anchor number.

9.10.2. Framing Bias

A framing bias is how a proposal or statement is put into perspective. Think about the last time you were at doctor's office discussing possible treatments. How does the doctor phrase the potential risks? The doctor always talks about the potential for success, even if that number is lower than that of a potential failure. Assume that the surgery has a 30% success rate. That is the number you hear. You should also hear 70% fail rate. Framing the success rate gives a sense of positivity to the situation. The same goes for projects. When proposing an idea to further the success of the company, you offer up the positives. You frame the proposal as a good thing. If you are arguing against the project, you frame the negatives. It all depends on the risk posture of your organization, risk-averse or risk-seeking. Be aware of framing bias not only when you hear a message, but also when you deliver a message. Frame your idea for the best chance at success. Know your audience and their tendencies.

9.10.3. Sunk Cost Fallacy

The sunk cost fallacy is the idea that we spent money on it, so we have to use it. We let the history of time, effort, and spending contradict the performance-based decision. Eric Wright, the CEO of Vets2PM, described a sunk cost fallacy perfectly with his example of buying a ticket to a football game. One person has spent $40 on a ticket. Another person wants to go to the game, but there is torrential weather in the forecast. Both individuals would be risking their lives to attend the game. Wright asked who is more likely to attend the game? Most people in the survey answer the person who spent the money is more likely to attend. In reality, the answer is that neither of them is going to the game is more likely. The fact that one person spent money does not influence the chances of going to the game. The money has been spent already. It should not factor into the future decision. If your life or the project's life becomes affected negatively, the decision should be obvious.

9.10.4. Regression to the Mean

Regression to the mean relates to streaks or the "next big thing." Systems have a tendency to regress to the mean or average. Consider the stock market. Some weeks are better than others, but the S&P 500 continues to grow. If you react when it is bad (sell) or when it is good (buy), you are playing with fire. As I have discussed in this book, our ability to predict is quite awful. Our projects are not an exception. Estimates are provided as the mean for the project. When a storm sewer item is bid, we bid it for a mean production. We do not factor in rain days, slow or fast production, or any other events. The 24-foot average for the 30-inch pipe is an average production we expect for the project duration. When we have days of 30 feet, we expect to have days of 18 feet to regress to the mean. Knowing this will prevent overreaction to good or bad days. If the good days continue, the project will be a success. If the bad days continue, we learn and bid the next project accordingly. Emotions also tend to regress toward the mean. Each person has an emotional baseline stance. We get excited, angry, sad, and so on, veering from baseline, but we normally regress to our mean emotional baseline. Keep this in mind when an employee or manager has a bad day. You might not be the problem, but you may be blamed for it.

9.10.5. Hot Hand Fallacy

The hot hand fallacy refers to someone being on a hot streak, with the expectation that his or her chances of success are different because of it. Gambling and basketball come to mind. Ever sit down at a blackjack table, and everyone is winning? It does not matter what the dealer does; the cards keep coming up winners. People start to bet more, play more hands, and people join in thinking they are going to strike it rich. Meanwhile, the table's odds have not moved one bit. Every hand remains the same percentage to win. This "hot table" idea can make someone go broke. The same goes for basketball. A player hits five shots in a row. Are the odds for the sixth shot different? Of course not. However, our tendency is to think so. We let past performance predict future results. The odds of making the shot have not changed. A streak should not influence our decision making. An estimator bidding projects and winning the last three does not guarantee that the fourth project bid will be awarded to that estimator. Treat each event independently to make the best decision.

9.10.6. Illusion of Control

The illusion of control is the overestimation of a person's ability to control events. Project managers have control over a project. But how much control do they have? Each individual on your team is exactly that, an individual. They all have their own mindsets, ideas, emotions, functions, and so on. To think you have control over them is absurd. You have a suggestion but not control. The same is true for project events. We have a budget, a schedule, and a scope. If we had control, none of those would ever change. We would always deliver projects on budget, on time, and within scope. How often does that happen? Our level of control is disillusioned with status, title, and ego. Realize that you can do everything perfectly, and the project fails. Your manager and senior managers do not want to hear about your illusion-of-control theory. This knowledge is for your personal mindset. Knowing you cannot control everything, and you have to leave things in the hands of your talented team, is a positive thought process and relieves the stress you put on yourself.

This section on biases is important to this chapter because we tend to beat ourselves up over poor decisions we make. Being aware of these biases can help us understand why we made that decision or why we did not. Is the sunk cost fallacy one you believe? Now that you are aware, you can take actions accordingly. You can explain to others why you made the decision when they start to question it. They too may believe the sunk cost fallacy. Each of these biases affects how we make decisions. Bringing them to the forefront allows us to identify them, think about the next step, and act appropriately.

Chapter 10 talks about public speaking and influencing your team. As a leader, you must be able to deliver a message effectively and convince your team. Without practice, these can be stress-inducing tasks. You will learn ways to deliver a message to your team better. Once you see your message getting across, this will make speaking to them easier. Chapter 10 talks about how people skim information and use it to communicate with them. How often do you send emails with long blocks of text? No one reads those emails. Actions speak louder than words. A phrase often said but how often acted? Is it more "Do as I say, not as I do"? I talk about ways to increase your influence, office politics, and get back on your horse.

Chapter 10

Speaking and Influencing

Delivering a message to a group is difficult. There is a lot of information and not a lot of time. Part of that message needs to be entertaining. When I was younger, I vividly remember my grandmother reading *The Wizard of Oz* and changing her voice for each of the characters. It was my favorite book for her to read because it was the most entertaining to hear. The changing of voices led to the book coming alive. Its message was more pronounced because I was engaged. If she had told that story in a monotone voice, I would have chosen another book for her to read. The same goes for delivering a message to your team. Use emphasis, inflection, drama, and so on, to enhance your message. Telling the same story and using the same tone will have your team daydreaming. Once attention is lost, it is difficult to regain.

When I was a laborer, there would be days I did not feel like working. It might be 90 degrees and humid. There would be plenty of manhole adjustments, and the day could not end soon enough. The crew member I worked with would start the day saying (jokingly), "Get your mind right." It was a way for us to prepare for the day ahead yet keep it lighthearted so we would not go insane. Mixing mortar and carrying heavy concrete rings is no one's idea of fun, but we would attempt to create the most fun possible. It was us two against the adjustment world. "Keep your head up" was another "joke" we would say to each other. Both of these sayings, while throwaway lines as they are, had a way of influencing us. They would put a slight smile on our faces and keep the day moving forward. Having a "motto," as we did, can help people get through the difficult times and the drudgery part of work. Keep the mood light in these times and even develop inside jokes of your own that team members can latch onto when frustrated. For us, it was all we needed during those times of difficulty and frustration.

Mental visualization was a technique my high school basketball coach would use before, during, and after games. He wanted each of us to imagine how the game would play out. The idea was to go through the motions in our heads, so when it came to the game, we already had a plan. He would constantly drill us on the "perfect possession." In his mind, the ball should never hit the ground even if the other team would score. "Grab the ball out of the net" was his constant reminder to us of his vision. His idea of perfection was to rebound the basketball, pass it to the point guard, get the ball to a wing player in stride, then pass it to a player running to the rim for a layup. At no point during that scenario would the ball hit the floor. He would stress to us that the pass is quicker than the dribble. It was also a way for us to forget about a basket we had given up and an attempt to score quickly on the other end to negate giving up the points. If the initial fast break did not work, we had a secondary break designed for a quick score. His objective for us was to be constantly looking to score as many points as quickly as possible. His ability to drill on those aspects continually made our team so much better. Going into each and every possession with a plan helped our team succeed more often than not. Influence your team. Become that person to give people a plan to follow daily. You should have checklists available. There should never be a lull for your team. If there is nothing physically to do, have them work on their mentality, positive or negative. People always have room for improvement.

When standing in front of the room waiting to make your speech, you are the general of an army. What you say, goes. They make one wrong move, you take them down. You must command that room with a presence and confidence not seen before. In my experience watching people present, it sometimes appears as if they are acting like a general instead of being one. It is obvious who has presence and who does not. The issue I have with that is this skill can be learned and crafted. Public speaking is not a mystic art form that only the most heavenly possess. It is a muscle that can be worked out and honed. Instead of looking silly in front of people a handful of times and then become a great public speaker, people would rather struggle through it and act as if nothing was wrong. I remember sitting in a large gathering of middle and senior managers. In front of us was the president of the company delivering a speech on how we need to work together. He was reading off of a teleprompter, a common practice for speeches. The issue was that he continued to stumble through words, skip over parts, or repeat things he had already said. Since it was my first time hearing him speak, I thought it was nerves. While talking with others after his speech, it became apparent that this was the common theme in his speeches. My reason for being harsh about this individual is that he was the *president* of the organization, not a middle manager speaking in front of his bosses. He was THE boss but could easily have been confused for a middle manager because of his delivery and lack of confidence on stage. Do you want that person leading you? I want someone up there who commands my attention, sounds like he or she knows what he or she is talking about, and delivers the message with a passion. Instead, I got a speech written by who knows and delivered with a whisper, not a roar. When talking to your team, be cognizant of how you sound and look while in front of them. Be confident. If you do not know something, tell them you will look into it and follow through later. Words do not mean anything; actions do.

10.1. Stoicism

> *Many lumps of incense on the same altar. One crumbles now,*
> *one later, but it makes no difference.*
>
> – Marcus Aurelius, *Meditations*

It does not matter where or when the mistakes happen. Get up, dust yourself off, get back on the horse, and keep riding. Have the influence over your team to give them that mindset. Create an atmosphere in which mistakes are encouraged and viewed as learning opportunities. Whether your mistakes come early in your career or later, you are going to make them. It does not make a difference. A senior manager can feel pressure always to have the answers and to be correct. That mindset is dangerous. People should constantly be learning and should never feel they have all the answers all the time. At one point or another, the incense will crumble. If you realize that, bouncing back should not be necessary because you will not internalize the failure. If you think your experience and knowledge are impenetrable, then your fall is going to hurt. Embrace the Stoic mindset that you are going to fail. Nature does not care how old you are or how long you have been doing this. Mistakes will be made. It is how you come back that makes the difference in a great leader and influencer. Admitting your mistake to the team lets them know that even the project manager can make a mistake and take responsibility for it. That approach will influence them to take risks necessary for success and take failure as part of the game.

I like to use the analogy of a "closer" in baseball. Every time this pitcher takes the mound, he is expected to get those last three outs of the game. It is his one and only job. His team has played eight innings and given him the lead so he can bring home the win. If he blows the save and the team loses, the loss is all on his shoulders. He has to face the music in the locker room from his teammates and the press. He feels like he took all the hard work his teammates provided and flushed it away because the result was a loss. But still, he has to go out there the next day and do the same job over again. If he allows that failure to enter his mind, he is going to let the other team win again. You must be able to block out the past performance and focus on the present performance. What is done is done. There is no going back. Now is the time to make up for whatever misfortunes you may have had. Even the greatest closers in the game have blown a save, but that did not deter them from becoming the best; much is the same for project managers. Do not let one day affect the next, success or failure. Instill that belief in your team.

10.2. Taoism

> *Therefore in the Sage's desire to be above the people,*
> *He must in his speech be below them.*
> *And in his desire to be at the front of the people,*
> *He must in his person be behind them.*

> – Lao Tzu, *Te-Tao Ching*

Lao Tzu gives a reason for why rivers are kings of their valleys: because they are good at staying below them. As a leader of your team, you have to be aware of how they perceive you. This Taoist philosophy concerns your appearance as a leader. While being the leader, you have to be one of them. In today's professional golf, the younger players are referring to their player/caddy relationship as "we." "We" decided to take driver on the short par 4. "We" decided to lay up on the par 5. While the player is the only one making the shot, the player is still referring to the team as "we." This example is the leader falling back in line—below the caddy the golfer leads. When push comes to shove, you have to take responsibility as the leader, but when things are going as planned, you should be blending in with your team because you are one with them, not standing out in front shouting from the mountain top.

Recently, I witnessed a conversation between a leader and an employee. From the outset, the leader was confrontational and established himself as the leader. He wanted an argument. How do you think that makes the employee feel when the leader intentionally wants to start an argument? The situation leading up to this encounter revolved around a meeting we had three months prior. Every winter, we gather up the foremen to go over the mindset for the upcoming season. We discuss any changes to policies or laws. We cover material that is required by our company regarding safety, workmanship, human resource issues, and so on. At this meeting, we emphasized how important it is for the foremen to enter their time, equipment, fuel, and production.

Fast forward to the beginning of the construction season and some foremen have not been doing it. The conversation I witnessed was the leader asking the foreman, "What legitimate excuse do you have for not getting your time in?" The foreman responded, "I usually put the information in on Sunday night." Immediately, the leader responded, "That is not my question! What excuse do you have for not entering your time?" The leader then entered a state of talking down to the foreman. This conversation is the opposite approach to what an entrepreneurial project manager should take. You have to remind the foreman of the talk you had in the winter meeting. Emphasize the importance of getting the information entered daily. Let the foreman know that you understand it can be difficult if there are issues to work on in the field, but it is important to enter the information. If the foreman cannot enter the information for some reason, he should be sure to call someone who can. Instead, in this example the leader chose to dominate the encounter. How do you feel when someone is talking down to you? Does communicating this way motivate you to do what the leader is asking? Always be aware of how you appear to others. Words and tone are important in any conversation. Speak on the same level, or, as the quote mentions, a level below the other person. A one-way communication channel is not a conversation; it is preaching.

Discussing difficult, intricate concepts at a level beneath the person you are speaking to is another way to implement this Taoist philosophy. Projects are complicated. There are many levels to them, many people involved, and a lot of money at stake. Any misunderstanding along the way will affect the outcome, most likely in a negative way. I was watching a television show, and there was an issue with accounting. The manager had the accountant come into his office and explain it to him at an eighth-grade level and then explain it to him at a third-grade level. It was for comedic effect, but the principle applies. Explain things at a level beneath the understanding of the audience. In dealing with sewer and dirt foremen, they understand what they do. The thought of tracking subcontractors and change orders is not a priority. They deal in production. If the pipe is going in the ground and dirt is being moved, they are satisfied. It is my job to get them to understand the importance of tracking the production. That is the linchpin to getting paid. Without money, there is no pipe to put in the ground and no more dirt to move.

Explaining this to them at the level of organizational expectations will not hit home. They probably do not understand—or, frankly, care—about the accounting department and using the correct purchase orders or job numbers. That is something people in the office do. They are field personnel. They get things done. As a project manager, it is important to explain this concept on a level they will understand, such as the importance of keeping track of change orders so we get paid for the work we perform. Not just getting the work done at a rate faster than we bid, but also that we receive the money for work performed outside the scope of the project. While the task is basically the same (installing more pipe or excavating more dirt), it gets applied to a different cost code because it is outside the scope. To explain that to field personnel at a management level will not be effective. Relate it to a hobby or activity they do outside of work. Would they want to perform a task for free? I doubt it. So why should they expect that from their organization?

This application can even work for senior or executive managers in explaining to them the importance of taking on a project. Use language at a level below their expertise. Make the presentation easy to understand, even for the most inexperienced of managers. Do not assume they know what you are talking about, as it is your area of expertise.

10.3. Sales

Grant Cardone has the motto, "Treat success as your duty, obligation, and responsibility, not as a choice or as a job!" Talk about influencing your team! What if everyone on your team had this attitude? Do not just go through the motions and do not stop at "good enough." Rather, ensure that your team members are improving themselves 1% or more each day they come into the office, and that they are doing their best each day. With that attitude, they do not have a choice but to succeed. This is the route you want your team to take. There is no room for negativity. They should focus on always helping each other, no matter the circumstance. Not only will this make your team stronger, it also will benefit your organization. Your team should be the one others look toward for guidance, and you should not be afraid to share your wisdom with others. Approach it in a helpful manner, and people will always be receptive.

The entire purpose of including sales in this book is to influence others to see the light you are shining. It is not to sell people an object, but an idea—a mindset. What is more powerful than changing someone's perspective for the greater good? That is teaching a man to fish, not just handing him a fish. He will continue to pass the knowledge on because of the impact it had on him.

This is the true difference between an amateur and a master. Making success your duty is something a professional does. An amateur wants what a professional has without taking on the personal responsibility for making it happen. Professionals consistently obtain results; luck is not a factor. Amateurs rely on luck and blame luck for their misfortunes. Professionals laugh at luck, knowing full well the system they have developed got them to where they are. Create a crystal ball for yourself so that you know how people are going to react or what they are going to say—is what a professional does. It means years of studying, not only someone else's work, but also your own. In Chapter 9, I mentioned taking lessons learned from both successes and failures. Those notes are going to build your crystal ball. You will develop phrases that will elicit emotions from team members or clients. It will seem as though you are magical when it is your attention to detail that is giving you access to that information. People will tell you all you need to know; you just have to listen.

Think of the influence you will have when you know what people are going to say or how they are going to react.

Small wins are so important but rarely are celebrated. People are always after the big score, the million-dollar project, or the 50% raise. Influence your team with small wins. For example, think of having a day when you have a paid lunch. In my laboring days, one noon hour the superintendent bought pizza for lunch along with two-liter bottles of soda, and it was as if it was Christmas in July. Everyone sat around for a half-hour eating pizza and enjoying the free lunch. That small win boosted morale more than the most productive day ever could. The same works for free t-shirts, vests, hats, jackets, or whatever. Suppliers drive up to the job site and start handing out hats; it is as if the team has never seen a hat before. They throw it on with the biggest smile, just as if they had won the lottery. Think in terms of numbers, not magnitude. In 2016, the Golden State Warriors of the National Basketball Association won 73 games, a new regular-season record for the most wins. One of those losses came to the Milwaukee Bucks (my favorite team), who ended up with 33 wins. The magnitude of that Bucks win was greater than most of the Warriors' 73 wins, but wouldn't it be better if yours was the team with 73 wins? The point is to rack up victories. Beat the teams you are supposed to beat. Reach the goals you are supposed to reach. Do not get caught up in the largest goal possible while leaving others behind. Collect the other goals on your way to the big one. People can lose sight of the end result and get frustrated; therefore, give your team little goals along the way. Then celebrate when those smaller goals are reached.

Earlier I mentioned the dog I have and how she has taught me to be a better project manager. Little wins with her are an excellent example of this. Teaching a puppy to sit and stay is a large task. My puppy often became so excited she would forget the task at hand. So the smallest, shortest sits would get rewarded, to bring the puppy back to reality. It seemed she was never going to sit fully, so the smallest attempts would earn her a treat. Before long, however, she was sitting and staying without my repeating myself. It is the same with your team. In trying to get your point across, reward the individuals making an effort, even if that effort is not what you are expecting. Eventually, rewarding those small victories will have your team performing that task without thinking and without the reward, because the next goal is more enticing. Without those rewards along the way, your team will become frustrated with the new way of doing things and stop doing it, or perform so poorly you will not ask it of them again. I saw this time and time again.

A company I worked for would try to implement a new way of tracking production. The foremen would get frustrated with the new system and stop using until the weekend. We wanted them tracking their performance during the week so we could monitor it better and make the necessary changes. Without rewarding them, their frustration won them over. Senior management's way of dealing with this was to call them and chew them a new one, further increasing their frustration. We are human after all. We like being rewarded for a good job. To clarify what I mean by "reward," I mean saying "nice job" or the equivalent. I am not suggesting paying out bonuses or bringing in lunch every time someone does something right. I am suggesting pointing out to the group a job well done.

10.4. Skimming Information

Electronic communication is increasing by the hour. Email inboxes are continuing to fill. Phone calls remain important, but for day-to-day communication, especially with the increased numbers of people working away from the office and even out of the country, email is the main driver of communication. People are required to read the information you are trying to get across rather than hear the information. The techniques for listening and reading differ.

With listening, suspending judgment and remaining curious helps to translate the message from someone you are speaking to in person. Suspending judgment of the person giving you are the message and also your assumptions about the message being given to you. How often has *that* person come up to you talking about work? From experience, you know gossip is the only thing on this person's mind. But on the off chance that they have something to say, you listen to them. Every so often, the message will be important and of use to you. Suspend your judgment until they begin to talk. Curiosity will keep you involved with a conversation. If you want to hear what someone is saying, what he or she says will translate better than someone talking from whom you have heard it all before. Body language also plays a role in face-to-face communication. How you act and speak plays a role in the receiver's ability to translate the message, which again means that you need to be aware of how you come across to others.

Those techniques do not apply to the written word. You do not have to worry about your posture when reading an email, or read every word as if it is your last. In all likelihood, you will skim the email to find the parts that interest you. Whether it is a start date, arrival time, or yay or nay, you want to seek out those points and then reply. Not only do you use this approach, your team members and stakeholders also do. The Nielson Norman Group (Nielson, 2006), using

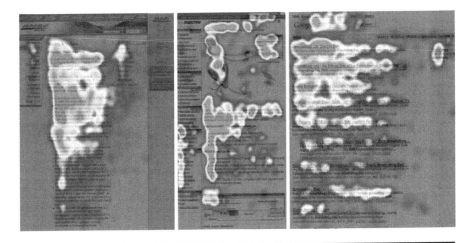

Figure 10.1 Heat maps from eye-tracking software of people skimming Internet articles. The dark areas indicate where readers' eyes focused the most, while the light areas indicate their least focus.

eye-tracking software and heat maps, have shown that this skimming pattern resembles an "F" (Figure 10.1).

The following are ways to communicate effectively with your team and with stakeholders under the assumption that they are skimming for information.

- Keep your message short. If you do have to type a longer-form email, keep in mind the "F" pattern of skimming. Format your email appropriately.
- Use humor to entertain. Memes and GIFs can create curiosity in the audience to keep them engaged. Make sure the images and content are appropriate for work.
- Use multiple media for the same message. Some people learn by doing, while others can read an article and instantly acquire the information. Videos may be an alternative to show your team how something should be done.
- Use microlearning. This technique incorporates short bursts of information requiring the receiver to demonstrate the knowledge back to you. For longer meetings, this technique will keep your team engaged for longer periods of time, as there is pressure to perform throughout the meeting.

Apply these techniques to your learning style as well as the way you communicate with your team. Realize that they are not reading every word. Be aware of your email formatting. Is it easy to read? Are there breaks to give the reader's eyes a rest? The information can be important and useful, but if no one reads it, the effectiveness is lost.

10.5. Speak Through Your Work

How many times have you heard, "Actions speak louder than words"? Do you believe it to be true? Saying you are going to perform and actually performing could not be further apart. Your team will catch on quickly. You will lose any influence gained previously. The team will become restless and bored, causing them to become increasingly passive. The team may even regress to norming. Your actions do speak louder than words. If the team sees you stepping out of your comfort zone, it becomes easier to ask them to do the same. Feeling motivated and energized, we can overcome almost everything.

Think of a time when you were delivering a message to your team when you did not believe in that message. How were your words being said? Were they enthusiastic? Did you show passion? What was your body language saying? All of these factors have an influence on your team's ability to accomplish and surpass goals. If you do not believe, who will? They take their lead from you. If they see you remaining challenged and needing their help with this new mission, they will respond far better than if you go through the motions of the next project. Remember, people see you from the outside. Their viewpoints and perspectives are never how you imagine yourself to be. They do not know the emotions and feelings you have. The team will base their enthusiasm up how the message is being delivered, and the actions taken after delivery.

There are many ways to speak through your work. Staying late is synonymous with caring about the results. It shows a dedication to your work. Including team members allows for chemistry to build and their contributions to enhance the time you are putting into your work. Your team wants to help, so allow them to do so. Clearing a hurdle as an individual is great, but clearing that same hurdle as a team goes much farther in terms of the greater good of the team. Do not just include them; make sure the task is worthy. Challenge your team. You will notice the difference immediately. Give your team the opportunity to spread their wings. Majestic bald eagles do not become that way staying in the nest. They need to get out, fail, succeed, and repeat until finally, they become great. Follow this approach with your team, and kick them out of the nest if you can.

Speaking through your work does not mean just mean taking action. Silence can be louder than any words spoken. Talking can be confused with getting things done. How many times have you been in a meeting resulting in square one? Nothing got done. People talked, took notes, stayed on schedule, but in the end, nothing was accomplished, yet people feel differently. Afterward, you hear people speaking of how great that meeting was and how all the topics were covered. Everyone goes back to their offices and continues with status quo. Ryan Holiday, the author of *Ego Is the Enemy*, writes, "We seem to think silence is a sign of weakness. That being ignored is tantamount to death. So we talk, talk, talk as though our life depends on it" (Holiday, 2016, p. 26). How many managers have you worked for who do exactly that? Because they are in charge based on their title, they feel the need to talk and talk and talk. All the talking usually comes full circle to their original sentence. If there was an editor for their transcript, the five-minute talk could have been summed up in one sentence.

Two meeting styles come to mind with all of this talk . . . talk. One meeting style is closed-door, no answering of cell phones, and stick to the meeting notes. If something comes up, leave to take the call, and we can discuss items that do not involve you; or wait until after the meeting is over and then return the call. Any interruptions had better be important; otherwise, you hear about it. The door is closed for a reason. These meetings may take one to two hours to complete. The second meeting style is much more relaxed: an open-door concept, all phone calls are answered, and interruptions are almost encouraged. Ideas are exchanged, and in the middle of the exchange, if a phone rings, it is answered. If someone stops by, there is a break in the action to allow for unfocused discussion. These meetings may take two to four hours, depending on the outside activity level. Which sounds more effective? Which gives the appearance that more work is being done? I say the first style is more effective, but the second style feels as if more is getting done.

The first style, a focused, idea-centric meeting, is the most effective. The issues and solutions are the focus. There is no time to answer a phone call from the shop or have office staff drop off invoices. This time was set aside to go over the plan for future projects. The talk is kept to a minimum. Once the issues are handled, then talk about the weekend can occur; it is not a "no-fun zone." Professionalism, however, is of the utmost importance. The second style allows for more talking, which gives the impression of more getting done. More words must mean more information and ideas. In actuality, the words are too often filler and fluff. It is like a description of the weather in a book: Unless it has a purpose to the plot, it is unnecessary. Those words may fill up the page, but the reader does not go anywhere. Losing weight comes to mind with this economy of words. How many times have you heard someone say he or she is going to start a diet? Months later, you hear excuses as to why it has not started or why it is not working. People feel a sense of accomplishment when they talk about doing something.

10.6. Increase Your Influence

This chapter is not only about speaking in front of your team and delivering a message that sticks, but also influencing them to greater heights. In *The 21*

Irrefutable Laws of Leadership (Maxwell, 1998), John Maxwell talks about ways to influence your team through leadership. One of those laws is "the law of the lid." The "lid" refers to your ceiling as a leader. If you rate yourself a 5 on a scale of 1 to . . . , the greatest height your team will reach is a 5. It is similar to gymnastics. A routine has a maximum score. A gymnast cannot score any higher than the difficulty determines, just as you cannot score any higher than your rating. This rating determines how effective you feel as a leader, and leadership ability determines a person's level of effectiveness. They go hand in hand. There is no separation of the two. So how can we grow our lid? Here are two ways:

- *Help yourself.* Continuing education is important to expanding your lid. As you continue to learn, your lid will grow. Watching webinars, reading books, and listening to your team are all ways to help yourself. Ryan Holiday says you should not only be a student, but remain a student (Holiday, 2016). With those words, you will have no problems growing your lid and becoming the best leader possible.
- *Add value to others.* Helping out others with their problems will, in turn, help you out with yours. If you are constantly a one-way street (people only help you), your team will get tired of only helping you. When they need some help, you tell them to figure it out themselves. This is a lazy way of leading. The goal is to have constant communication and develop a street going in both directions. You help them; they help you. You add value to their lives; they add value to yours. This give-and-take approach will increase your influence on your team.

A second law John Maxwell talks about is "the law of influence." As a leader, you must have followers. Without followers, you are not a leader. Maxwell presents three questions. If you can answer these three questions, you will always have followers:

- Do you care for me?
- Can you help me?
- Can I trust you? (Maxwell, 1998)

Let us answer these in order of difficulty. The easiest to answer is "Can you help me?" As a project manager, you have answers. Those answers help people. Project managers always have the ability to help. "Do you care for me?" is the middle difficulty. Are you asking questions such as "How was your weekend?" or "How are the kids doing?" or "Did you get any deer, turkey, etc.?" If not, your team may question whether you care about them. These are simple questions to ask, but they show you have an interest in their lives outside of work. We can get caught up in the daily struggles of the project without taking a second to recognize the human side of our team. Showing you care goes a long way. I remember someone asking me, "How are you?" with a sincerity that stuck out to me. Most of the time that question is asked in passing, but this time, it was asked with a caring tone. It has affected me to this day because it showed a level of caring I had not seen in the workplace before. Lastly, "Can I trust you?" is the most difficult to answer. You need to build that trust within your team. If what you are saying comes off as hyperbole or straight-up lies, your influence takes a steep drop. I recall managers lying directly to my face about some small issue. If they are going to lie about a nonconsequential issue, think about the extent of future lies. Do not make your statements a guessing game among your team. Give them the truth and nothing but the truth.

In Chapter 4, I discussed storytelling to help execute your strategy on an organizational level. Storytelling can be a great way to influence your team as well. Storytelling creates access to the heart. Josh Waitzkin, the author of *The Art of Learning*, wrote, "He had to teach me to be more disciplined without dampening my love for chess or suppressing my natural voice" (Waitzkin, 2007, p. 9). A project manager must tap into what makes his or her team tick. Notice the fine line Waitzkin's coach had to walk to get the best out of him. Waitzkin was a natural, crossing over to traditional coaching of chess. He wanted to broaden his knowledge of the game while still fueling that natural passion. Your team may be similar. How can you take their excitement about the new project and carry that over to getting things done? I had a professor in college who was nicknamed "Dr. C," because that is the grade you got in his class. He was proud of that nickname. He set the tone of the class on day one with that little story. He was going to be tough on you, and if you came away with more than a C, you are a good writer. I ended up with a C, as did 90% of the other students. That story influenced my performance in the class. If my grade before the class even started was going to be a C, you are going to have a tough time motivating me to do better. The professor's mind was already made up. Do not be a Dr. C. Remain tough on your team, but do not influence them toward mediocre performance before the project even starts.

10.7. Office Politics

Office politics. They exist. They influence. How can project managers use them to their benefit? Bruce Harpham's webinar, "How to Play Office Politics Without Losing Sleep" (Harpham, 2016), offers fantastic ways to play the game without compromising your value system. Harpham gives three strategies you can lever-age: decision-making biases, engage your boss, and do your public relations.

The decision-making bias I want to discuss is confirmation bias. I discussed it in Chapter 9, but not in the context of office politics. People have different reasons for deciding which projects are best for the organization. As a project manager, you must learn their reasons and use confirmation bias to assist their decision making in your favor. A computer program can do many things. The accounting department will have a different interest in its ability than the esti-mating department. When introducing new software, you have to play to the team's confirmation biases. They want to hear why it works for them, and not why it works for other departments. If it works so well for estimating, why does accounting have to implement it as well? People may ask why does the program not just stay with estimating? You have to find a way to relate your interests in the new program to all departments.

Engage your manager. Harpham talks about learning your manager's priori-ties and style. I want to focus on style. In quarterly reviews, I used to get poor ratings on my working habits because they did not fit with my manager's work-ing habits. I wanted to get the tasks done and in his hands as quickly as possible. The issue came with minor edits to the documents. There would be a missing word here or a line in the notes I would forget. He wanted me to take my time and make sure it was correct before bringing it to his office. Our working styles did not match up, and it caused us both frustrations. My strategy was getting it to him 95% complete, and he could make any necessary adjustments. His strategy was for me to get it to him 100% complete, making the minor adjust-ments on my own. We did not collaborate very well together. If I had learned his style sooner, our working relationship might have been better. Making your manager's job easier allows for you to play the office politics game better. Your manager has a say with people who matter.

Finally, do your public relations. In Chapter 3, I talked about being the "white rabbit" of your organization, someone who stands out from the crowd. This is a great way of doing your public relations, but Harpham cautions, "the dose makes the difference between medicine and poison" (Harpham, 2016). If you start to sound arrogant rather than confident, the tables will turn. Your influence will turn to dissent. People will turn against you because your brash-ness is a turn-off. You want to stand out without overshadowing your team. Earlier in this chapter, I spoke of letting your work speak for itself. This way of promoting yourself is far better than talking. If you are performing at top levels, it will show without your bringing more attention to it. Performance reviews are the times to speak on your behalf about the work you are doing. It is your time to shine behind closed doors with the people you need to influence the most. Any other setting does not have as much impact as those performance reviews.

10.8. Getting Back on the Horse

Project managers fail. Schedules get behind; plans go out the window; stakeholders become upset, and project managers do not have the answers. The situation is stressful. Self-doubt creeps into your mind. Questions about the future pop up. How can project managers get back on the horse? Bill Walsh, the legendary football coach of the San Francisco '49ers, has five do's and don'ts when faced with adversity. Here are the five do's:

- Do expect defeat.
- Do force yourself to stop looking backward and dwelling on the professional "train wreck" you have just been in.
- Do allow yourself appropriate recovery—grieving—time.
- Do tell yourself, "I am going to stand and fight again," with the knowledge that often when things are at their worst, you are closer to success than you think.
- Do begin planning for your next serious encounter. (Walsh and Walsh, 2009)

Self-doubt is not a theme anywhere in these five key points. Walsh wants you to take the time to feel and then start up again. Failure is not a time to reflect on the multitude of mishaps you have had. It is a time for learning. There is positive self-talk going on throughout the failure process, and it is a process. The first bullet point tells you to expect defeat. How often do we envision ourselves failing? This question is not to say you are going to fail with every opportunity you take on, but the possibility of failure should be there.

Walsh also suggests five don'ts after you have failed and need to push on again. Here are the five don'ts:

- Don't ask, "Why me?"
- Don't expect sympathy.
- Don't bellyache.
- Don't keep accepting condolences.
- Don't blame others. (Walsh and Walsh, 2009)

When you fail, there is often a feeling of "why me?" or self-pity. You look for others to blame. You get down on yourself. During this time, you are failing to realize that this happens to everyone. Your manager fails. Your team fails. Your stakeholders fail. Most important, you fail. There are no excuses for failing. Again, take these opportunities and learn from them. Continuing to accept condolences means you are still thinking about the failure and have not grieved properly.

As Chapter 7 discusses the pessimistic side of project management, these do's and don'ts come off as very pessimistic. Expect to fail? Plan for the next one? There is going to be more failure ahead?! Don't bellyache? All of these come off as callous, but they are necessary words when you need to get back on the horse and start riding again. Most of us have been yelled at and chewed out by a manager. Did that stop us from pursuing our goals? I doubt it. That is how these mindset triggers should work for you. Feel the feels. These emotions are good learning tools. Take them with you the next time you attempt something great. Eventually, failure turns to success; then you will look back on your failures fondly.

Chapter 11 takes a contradictory approach to "It is what it is." I have used this phrase, usually after making a decision that does not go as planned. The team must adjust, and we move ahead. This chapter discusses situations that are not as black and white. There is nuance throughout project management, where the lines become blurred. Science turns to art. I talk about how two people can look at the same situation and come to different conclusions. The transitive property makes an appearance. Days of high school math come rushing back. It classifies how two things can be equal on two separate occasions but not make everything in the situation equal. An A does not equal a C. Alchemy is discussed, along with the power to transform a common metal (or idea) into gold. You are taking something that is not, and making something that is. This approach takes a great deal of social intelligence.

Chapter 11

It Is What It Is Not

I attended a project managers' meeting at Discovery World. It was after hours, and we had a private tour of the facility. When we approached the stingray tank, there was obviously something out of place. In the tank, one of the stingrays was completely black, while the others blended in with the surroundings. The tour guide explained that this was due to a puffer fish incident, as a result of which the stingray had become blind. Stingrays have a similar ability to chameleons: They can change their color based on their surroundings. Because Flipper the stingray was blind, she thought her surroundings were black. I found this to be a valuable lesson in perception. The other stingrays must have thought she was crazy for sticking out like that, but in her world, she was completely safe and camouflaged.

Social media has given life grandiosity. People are constantly posting vacation pictures, the best versions of themselves, and so on. This viewpoint can skew people's perspective on their lives. Media portrays the hero without showing the hard work and dedication they put into their craft. Steph Curry, often regarded as the best shooter in the NBA, won the 2015 National Basketball Association Championship, the 2015 and 2016 Most Valuable Player awards, and is a three-time All-Star (2014–2016). He puts on a dribbling and shooting clinic before and during games. No one sees the hours and hours and hours of practice put into dribbling and shooting, and no one sees the conditioning it takes to play at a high level each night. People take for granted the behind-the-scenes work because it is never shown. The same goes for the project manager who continues to influence and produce and lead. No one sees the project manager reading books, drawing up plans and schedules, or the mental games being played back and forth during negotiations or when trying to pitch senior management to get the lead on a project. People do not see the failures that have led up to the successes. Classic statistics belong to Brett Favre, Hall of Fame National Football League quarterback: He threw the all-time most touchdown passes but also the all-time most interceptions. While the touchdown passes are celebrated, the interceptions are tossed aside. Do not fall into the trap of believing that some people are just luckier than others. Hard work is required to get to the top. Having an influence over others as well as yourself is required.

In *We Learn Nothing*, Tim Kreider discusses a decision as coming down to "middle school politics" (Kreider, 2012). He was speaking with a friend about peak oil, the hypothetical point in time when the global production of oil will reach its maximum rate, after which production will gradually decline. His buddy was preparing for the worst by setting up a bunker and preparing for riots on the streets. When thinking about peak oil and his own future, Kreider was asking himself questions, but not ones he was expecting: ". . . the most relevant questions turned out to be not 'Does the evidence support this theory?' or even 'Is Ken trustworthy?' but 'Would I rather live in the peak-oil compound with Ken or die in the food riots with Harold?'"(Kreider, 2012, p. 128). The facts were not questioned, but his future comfort was. This realization is important when applied to negotiations. Sometimes the issue is not whether you have a valid point, but whether your project or your salary increase helps the manager you are trying to convince. When you prove you are a problem-seeking and solutions-oriented person, the price goes out the window. If your project can solve a problem the organization or the manager has, it will get approved regardless of the financials. It does come down to middle-school politics. If the manager or organizational leaders have taken a liking to you, your odds of success increase. People want to help their friends. Do not perceive the situation any differently. As we have learned earlier in this book, there is a price for complexity. Sometimes a decision boils down to being liked or not.

"Outrage is healthy to the extent that it causes us to act against injustice, just as pain is when it causes us to avoid bodily harm," wrote Tim Kreider in *We Learn Nothing* (Kreider, 2012, p. 53). People who are quick to judgment and create an uproar for no apparent reason must be handled immediately. Think of social justice warriors who will back a cause because everybody else is doing it. People will follow the loudest voice and not begin to question the reasoning. There is a popular "experiment" going around social media that showcases this behavior quite well. I will label it a fable rather than give it scientific credit as some will nonsensically do. The story goes as follows:

Scientists placed monkeys into a cage with a ladder. Atop the ladder, bananas were placed. Each time a monkey climbed the ladder and obtained a banana, the other monkeys were sprayed with cold water. After a period during which one monkey got a banana while the others were sprayed with water, the monkeys who were getting wet started to beat up on the monkeys trying to get the bananas. Soon after, no monkeys were going up for the bananas, because they did not want to get beaten up for their actions. The scientists then replaced one of the monkeys with a monkey that was not originally involved. The newest monkey would see the bananas and begin to climb the ladder, only to get beaten up for attempting to get the bananas. This monkey had no prior experience with water being doused on the monkeys. The monkeys continued to be replaced one by one until the cage was filled with monkeys who had never actually been exposed to the water treatment. The newer monkeys would beat up any monkey attempting to climb the ladder, because they had been taught through experience that that is the way of the cage.

I include this story, not as fact, but as a lesson learned, the way children learn from fables or adults from fiction. While the story may be exaggerated or made up entirely, the lesson learned is very real.

Relate the monkey story to any new members of your organization or team. Have them stop and think about their actions. "Because that's how we do it" is not an acceptable answer. Make your team aware of why they fill out the paperwork in a certain way, why they meet at a certain time, why they must act and dress a certain way, and so on. Acting a certain way because others act that way is not a reason for doing something. Every action should have a purpose. Do not allow your team members to be lemmings. Mindlessly and aimlessly following should not be a goal of yours as a leader. You want criticism and push-back to challenge your ideas, to make them stronger. If you never have to justify your actions, you go unchecked, leading a team of "yes" men or women. Be strong in your values to allow for criticism, and do the same for your team members. Ask more of them, but do not assume they know the goals of your mission.

Anger is an initial emotion but may not be the root emotion. People become angry because that is a default emotion that has paid off in the past. Get to the reason why they are angry. It may result in a simple answer that can be resolved quickly. Having anger linger on your team is disastrous. Performance lowers. Teamwork lulls. How many times have you found out that a disagreement had arisen because of a poorly worded email? Tone does not translate well via text, especially if the team members are new to working with each other. If the individuals have a rapport, then email can be more effective because tone can be known rather than assumed. Be proactive in seeking out disagreements and solve them so the team can go back to performing. Not only does root-cause analysis work for project issues, but also for team member issues. The problem, more than likely, will not be as difficult as the team members make it seem.

11.1. Stoicism

The best revenge is not to be like that.

– Marcus Aurelius, *Meditations*

Once you figure out what style you do not want to incorporate, do not be that style. Instead of fighting fire with fire, as Bruce Lee says, "Be like water." Companies I have worked for struggle with this idea. They come from the field, and since they had to do it, you are going to do it. Always one-upping any horror story you have. If you cleaned out sanitary sewer manholes for 10 hours, they did it for 12. If you were knee deep in mud installing a pump, they were knee deep in mud during the winter. Instead of not being the things they despised in the field, they make you go through the same process for ego's sake. Why not be the change? Be the change sounds so simplistic, yet I rarely see someone trying to change the way things currently stand. If you see people being assigned tasks for the sake of delegating, do not take that approach. The classic line, "Be the change you want to see," works. It is so frustrating to hear people complain about a situation that they can change. If you are a project manager and remember how awful it is to clean out catch basins at the end of a project, build in the budget room for a vacuum truck. The time saved renting the truck will pay for the labor that would have occurred had you had two laborers responsible for cleaning them. Just because you did it does not mean it should always be done. Use your learning experience to make things better. Parents always want better lives for their children than what they had growing up. Take that approach to your position as project manager.

"Shoveling builds character" is another popular phrase I have heard over the years. A skid steer is vastly superior at moving dirt than any laborer in human history. Why would you threaten your success by making crew members shovel rather than use the equipment? This situation is a mindset I have too often seen in the construction industry. During a conference call about a wind farm project, there was an older person talking about how the newer-generation workers were soft and coddled, because they did not want to work out in the cold when it had been customary for the older people to do so when they were in that position in the company. If temperatures are below zero, nothing productive gets done because you spend more time warming up both yourself and the equipment than you spend working. This guy wanted those individuals to work in those conditions because that is what he had to do when he was laboring. Do not get your revenge by sending out people in brutally cold weather. Get your revenge by calling it a weather day and getting back to it when the conditions are more suitable for production.

11.2. Taoism

Thus with all things—some are increased by taking away;
While some are diminished by adding on.

– Lao Tzu, *Te-Tao Ching*

Addition by subtraction. How many times have you heard that, versus how many times have you experienced it? Use it when trying to find the root cause of a problem. By looking at what it is not and taking away possibilities, you are adding value to the discussion. When you have a disgruntled team member, and you remove the team member from the project, production may increase. In one year, four foremen were released and lost their jobs. To me, it was an insane number. To senior management, it was necessary. The show went on and did not skip a beat. Some of the plants produced more after the foremen were removed from the project. Sometimes an employee can be toxic not only to the organization but also to the team he or she is leading. An employee's attitude will start to spread like cancer and infect everyone it touches. If people are showing up already disgruntled, it is nearly impossible to change that on a daily basis. The project may be bid for five crew members, but if those five are working at half-capacity because one of them is disgruntled, removing that one employee could increase the capacity of the other four to full. If you recognize that employee for what the employee is not (a leader, motivator, producer, etc.), that is addition by subtraction at its finest.

Adding on to a project is not necessarily the best move. The phrase, "Too much of a good thing," comes to mind. If you sit down and eat an entire cake, you are going to feel sick afterward. No matter how delicious the cake may be, it is not meant to be eaten in one sitting. The same is the case with managers. If you keep adding on managers to manage the managers, that is too much management. A company I worked for had an entire division dedicated to managing projects, similar to a type of project management office. The difference is that the lines were blurred as to who was running the project. Was it the project management division? Was it the division that was the prime contractor on the project? Even within the divisions, there were conflicts as to who was working for whom. Managing projects is an obvious need, but having too much management is possible. There needs to be one voice making a decision; not one voice from each division that has a say in the decision but then picking one of the ideas and seeing if it works. In the example here, the organizational chart looked like a plate of spaghetti. The leader of one division was leading an initiative on removing of job titles, which made things worse. In effect, this situation was similar to taking that plate of spaghetti and dropping a bomb on it. If you do not know who to get a hold of, how will adding more management help? It was a problem, to say the least.

11.3. Sales

Do you ever purchase items without thinking about it? Because the brand or item has worked so well for many years, you continue to buy it without hesitation. That is the "second money" phenomenon. The first purchase breaks the ice; the succeeding purchases continue because the product or idea works so well. You want to build this rapport with your team and senior management. Some people get the idea to suggest as many options as possible to see if something sticks. That is fine if a brainstorming session is occurring. Otherwise, pick and choose your opportunities. Find out what your idea is and is not, present it, learn from it, and continue the feedback loop. After a while of presenting great, concrete ideas, your audience (team, stakeholders, management, etc.) will buy the thought without much thought. Your word will be trusted. Your freedoms will expand.

I have always been known as a quiet individual. I pride myself on that because I want my words to be heard, not just be another comment from the person in the corner. I am not the person to make things up and tell you what you want to hear. I have run into so many people who are quick to answer with words the audience wants to hear, and nothing gets accomplished because their intent is never clearly known. In a former company, we would always tell people we would be there in two weeks. It was our magic timeframe to give them a close enough date to satisfy them while not having an intent to get there in two weeks. It was a nice ballpark number to tell a client to keep them quiet for two weeks until they called again. I always had an issue with this game. "I don't know" seems to have a black cloud over it. It is rarely used, yet often people just do not know. Telling people two weeks with the intent of four to avoid confrontation leads to more issues. Sometimes you have to tell the client what the project is not. It is not going to happen in two weeks; it will likely be four or five. That honesty will go a long way, especially if more snags are hit along the way. Be forthright with people. Do not have them question whether you are telling them the truth, or something they want to hear. That is a lazy game. It is taking the path of least resistance and giving you a poor reputation even if people know your game.

Instead of people questioning their mental purchase of your information, your goal is to have them wanting more and coming back for seconds and thirds. We often think of selling as a physical exchange. We are discussing the mental selling of information. Let people know you know. Being a quiet person,

I often struggle with this concept. I want to play coy, sit back, and then dominate my small pond. I have to force myself to speak up with the knowledge I have and make sure people listen. It may be a struggle for you as well, but it will only make you a stronger project manager in the end. Knowing something and not sharing is worse than not knowing. It is more of a decision to know and not share than never to want to know or make an effort not to know.

11.4. Problem Managers—Problem Solvers

Problem solvers create explanations for why things happened and then produce a solution to resolve the issue. This situation can cause issues when unknown unknowns occur. Creating a narrative to fit your cause of the problem is a tricky task that needs careful consideration. Because something has happened one way for so long, do not rule out other possibilities. For instance, aggregate crushing equipment takes a lot of maintenance to stay running smoothly. Hard rock and quartzite are familiar reasons for wearing down parts of equipment. However, pointing immediately to those reasons rules out any other possibility for wear and tear on equipment. You are assuming the foreman and operator know how properly worn parts appear. You are assuming the equipment is being run correctly within its manufactured specifications. Ruling out human error could be a costly oversight. Including human error in your narrative for the problem creates a better picture of how to solve future issues. Even if a fishbone diagram is created, there are still parts missing because unknown unknowns are nearly impossible to consider. Pinpointing solutions should be as well thought out as the project charter, project plan, schedule, or any other part of the process. The solution is what it is not. Do not get caught up in past results predicting future outcomes. That narrative causes unseen problems without solutions. We, as project managers, have to be ready for anything. Fitting a narrative to a problem limits our creativity to solve future errors.

Think of a time when you looked at a spreadsheet, came to a conclusion, then someone else looked at that spreadsheet, and came to an opposite conclusion. It may not happen often, but ruling it out is naïve. Growing up, I would watch sports highlights continuously. Then people came on the television and talked about the games and tried to predict future outcomes. Never is a point clearer than watching two people argue over the same game as if they were watching two totally different games. One person thinks it was a foul. The other thinks it was not. One announcer wants him to get suspended; the other thinks it is not a big deal. The same can happen with data that are gathered up about a project. I was with a company that ran an organization-wide study on idle times of their equipment. It came back that 40% of the working hours were spent on idle time. That is an incredible statistic, and a small change could mean thousands of dollars back to the organization. The issue with the study was that the data set was incomplete. Only less than half of the fleet of equipment was studied. To make such a generalization without the full data set makes for a poor conclusion and even poorer solution. If the information you are given is a half-truth, then the solution is going to be half-hearted.

An issue is thinking linearly. If A happens, then B. If B, then C. And so on. This linear way of thinking takes the creativity out of problem solving and leaves large gaps for the unknown unknowns to occur, such as going back to past outcomes to predict future performance. Do not fall into that trap. Is a project ever black and white? Are specifications clearly stated? Never. There is always room for interpretation or nonlinear thinking. When was the last time a schedule was created and went as planned? The estimates and actuals rarely line up evenly, pointing to more evidence of nonlinear thinking being applied to project management. Expect the unexpected. Fast-tracking or crashing a project may be necessary to keep it on schedule. A linear thinker would never apply that logic. Construction linear thinkers are engineers. They design a project on a computer using GPS coordinates and contours from the region to develop a virtual lay of the land. When their interpretations are applied in the field, problems quickly arise. There may be utilities in the way of construction that were not planned. Soil conditions may differ greater than anticipated. In their minds, this was not possible before construction, but having seen the results, they need to adjust. Everything looks better on paper. The walls align perfectly with the foundation. The electrical and plumbing do not interfere with each other. In reality, people start working next to each other, and anything is possible. A worker can install something, thinking by doing so the project will be finished earlier, when later down the road that installation is in the way. Become a non-linear thinker when dealing with conflicts, whether it is schedule, resource, personal, or something else.

Silent evidence, a term I first learned from reading *Black Swan* by Nassim Nicholas Taleb (2007), is a concept similar to unknown unknowns. It comes from the fact that the winners write the history. The stories are told or written by people who have survived, whether it is physically surviving a war or mentally surviving your manager. It is the idea that strong evidence is often overlooked because the real evidence gets lost in the triumph. Examples are lost civilizations, unknown start dates, and reasons why something is not going correctly. A project manager can look at all the information available without coming to a conclusion because the evidence is hidden by the past. You only know what you know. That is a large limitation and often forgotten. An organization will continue to succeed in one way until it does not, and then its leaders will make changes based on why the present fails. Start to develop a proactive way to approach projects by seeking out the silent evidence.

An example is a search for good talent, whether it is to hire the person or have him or her join your team from within your organization. People will put a large price tag on talent because it usually leads to success far surpassing the money. The talent pool has not shrunk, but the ability to find the right people has diminished. There are so many talented individuals who want to work, but typically the people who are responsible for hiring are looking for the perfect employee or team member. Instead, they should be looking for the right fit. I tend to look at the New England Patriots or the San Antonio Spurs for inspiration. They will take a journeyman player, who has moved around to different teams his entire career, and turn him into a valuable piece for winning. If you had all superstars and the best talent, there would be unresolvable conflicts hourly. There has to be people willing to take out the trash, make coffee runs, and do any work required to help the team. They will complement the superstars and help them do their jobs to greater heights. The silent evidence is finding those people who will fit in your system, not necessarily those with the best résumé or the highest grades.

Diversity is a great way to introduce new viewpoints naturally. I worked for a company whose talent pool reach was limited. They hired similarly minded people to fill roles held by similarly minded people. There was never a shakeup in how a team operated. Divisions would lose money annually yet continue down the same road. Why not hire someone who has a different take on the industry or even life? Instead of being scared to take the chance, use the opportunity to learn. It may not work out, but you are going to learn something, either about yourself or about how to do things differently. Being comfortable is a bad thing. I see so many people who get comfortable doing the same thing over and over, some expecting different results (insanity come to mind?). Use these diverse skills to your advantage. Have a technical person along with a theoretical person work together. They will attack a problem totally differently and may even come to a solution that has never been conceived before.

In construction, I see so many technical people working together and never thinking about the theory behind things. Or they swear off technology because it always breaks on them. Introducing a theoretical person skilled in technology will change their lives for the better. I was that person for more than two years. It was rewarding for me to see one of those technicians using a computer effectively, even if it was a forwarded email that took no effort. It showed progress and a willingness to learn. It was frustrating to see people not use this ability to make their lives better because they did not want to learn anything new. This approach is where the change has to come from the top. Show them the capability it has and how you have used it to save time and make your life easier. View diversity as an advantage, not a burden.

11.5. A = B, B = C, Then A ≠ C

In mathematics, the transitive property is if A equals B and B equals C, then A equals C. In project management, this is not the case. Take analogous estimation, for example. Two projects seem similar on their surface. You provide a similar product to the owner at project completion. Why not apply the transitive property to the estimation and call it good? There are always subtle nuances within a project or with the owner that do not translate from project to project as easily as the numbers may suggest. Stakeholders' ideas of "complete" vary from project to project. Some are more involved than others. What lies beneath the surface makes the transitive property useless. Assumptions can be made to fit the transitive property and confirm ideas that may not be true.

Sports are a great example of why the transitive property does not apply. The 2007 New England Patriots had an undefeated regular season, going 16–0. They appeared unstoppable, and most people expected them to win the Super Bowl. No other team got in their way throughout their run to the Super Bowl. Heading into the game, the talk was not who was going to win, but by how many points would the Patriots win? People were blindly applying the idea that because no one had beaten them up to that point, no one was going to beat them. The New York Giants were their opponents in Super Bowl XLII. Analysts would compare common opponents to see how these teams would stack up against each other. One of those common opponents was the Washington Redskins. The Patriots routed them 52–7 in New England, while the Giants went 1–1 versus the Redskins that year. The Patriots and the Giants even played each other that year, resulting in a New England victory, 38–35. Since the Patriots were better against similar opponents and had even beaten the Giants earlier in the year, of course, New England would win and finish the season undefeated, right? WRONG. The New York Giants went on to win the Super Bowl 17–14, proving all the critics and analysts wrong. Life is too unpredictable to apply mathematical models strictly.

False analogies between two similar projects can cost organizations thousands if not millions. Construction provides daily examples. From one city block to the next, soils can change, underground utilities can vary, and neighbors can create unforeseen issues. Anytime a project involves performing work underground, expect the unexpected. A storm sewer can be installed at a record pace because soil conditions are prime, and underground utilities are absent. The next street over, soil conditions can be wet, with old utilities in the way of new construction. If you were to bid both projects similarly, having the easier of the two first, your production and schedule would take an immediate hit. The easier of the two projects had 30 linear feet of storm sewer per hour. Bidding the second project at that rate and only getting 20 linear feet per hour, you are losing money every day you perform that task. One would think the conditions would be similar because of their proximity.

There is no way of knowing absolutely, but you cannot assume based on previous experience. One way to test before construction is soil borings. You find where the deepest runs of a storm sewer are going to be and drill into the ground at those locations to see each location's soil conditions. Taking soil borings at many different locations will give you an idea of potential issues with poor conditions. They may not be a true sample of the conditions because of the small sample size, but it is better than assuming based on the prior project.

Visual inspections can help with assessing the lay of the land. There can be some surface clues that tip off underground conditions. If there are asphalt patches in the roadways that appear narrow, the soil conditions are good. However, if they are wide or vary in size within the same run of a storm sewer, there is potential for poor soil conditions. Being close to a body of water has the potential for wet conditions, as the water table may be higher the closer to water the project comes. These methods are not foolproof, but they can add value to the preproject assessment.

11.6. Alchemy

Anger is a powerful tool to transform old habits and replace them with new ones. Fear and sorrow inhibit actions; anger generates it. When you learn to make proper use of your anger, you can change fear and sorrow to anger, then turn anger to action. That's the body's secret of internal alchemy.

– Socrates, a character in *Way of the Peaceful Warrior*
(Millman, 2000)

Alchemy is defined as "a science that was used in the Middle Ages with the goal of changing ordinary metals into gold." A philosopher's stone was a substance thought to perform such a transformation. As a project manager, think of yourself as the philosopher's stone transforming ideas into successes. Be the substance to translate a brainstorming session into the next big thing. See what others do not. Be what others are not. Do as others will not. In the world of chemistry, the philosopher's stone has no place. There is no evidence to show ordinary metals turning into precious metals, specifically gold. However, in the world of project management, ordinary ideas can be turned into gold every day (Figure 11.1).

Figure 11.1 Transform your dollar idea into gold.

Take the quote above, for example. How many times during a project do you become angry? Something does not go your way, you lose a team member, resources have been taken away, and so on. Each of those would be a reason to become angry. Instead of using those events as excuses, use them as fuel to power your surge ahead. Show yourself and your team that nothing will stop you from trudging ahead and getting the job done. Turn what others deem as setbacks into creative ways to overcome.

Another example is dollar stores. In 1982, David and Shari Gold founded 99 Cent Only Stores retail chain. The concept is to sell a costly item for 99 cents to the first nine customers who line up on the opening day and other limited items for the next 90 new opening-day customers. With this model, 99 Cent Only Stores became a $650 million business. The dollar store concept takes a group of simple items, usually, which cost $1 per item and creates a great business out of these simple items. This success is your goal as a project manager with your team. How can you take a $1 idea and turn it into gold? What ways can you make the most of your situation?

11.7. Pride—Fool's Gold

If you can't swallow your pride, you can't lead.

– Genghis Khan

Project managers are leaders. We have a team to lead. We have an organization to help lead. We have ourselves to lead. All aspects of the position of project manager revolve around leadership. We become proud of the work we accomplish. Look at the infrastructure or software or community we have built. Isn't it great? What if the answer to that question is "No"? What happens when the teams and projects we lead are not as great as we think they are? Our pride becomes tainted. We have given ourselves to this mission, to keep stakeholders satisfied, and to maintain professionalism. As Ryan Holiday, the author of *Ego is the Enemy,* wrote, "Pride takes a minor accomplishment and makes it feel like a major one" (Holiday, 2016, p. 74). Because we are a part of the project, it feels grand to us. It is important because we are involved. If this is the case, the project manager is being given too much credit for the success or failure of the project. It is a team game. There is no one individual responsible for all things, bad or good, on a project.

I label pride as fool's gold because pride credits you before earning said credit. It gives you that feeling of accomplishment when nothing has been accomplished. This false sense of accomplishment harkens back to letting your work speak for itself. When pride does the talking, the cart is before the horse. Your identity relies on the success of the project. You are putting yourself out there so much that you are identifying with the outcome. Chapter 9 talked about how to keep from internalizing both successes and failures. Pride is a source of internalizing. Instead of continuing your path on improvement and trusting the system, pride says you are already there. You have done this, and there is nothing more to prove. Pride says you hit a triple when you were born on third base. It is calling you a leader because you are delivering a message in front of an audience. That is not what leadership entails. Leadership continues to foster relationships long after the spotlight is gone. It is giving advice in such a way that it connects instead of alienates. It is bringing together a team and performing at your best, day in and day out. Anyone can go in front of an audience, talk for an hour, and then go about his or her business. The challenge comes from getting your audience to take action on what you said. When your words have a lasting impact, it separates a leader from someone talking in front of people.

This approach is not to say eliminate pride. You should be proud of the title you hold or the successful projects you have built. The issue lies with the extension of that pride. New challenges are not sought after with the vigor you once had, because you have been there and done that before. This approach may result in complacency. Leading one project to success does not make you a project manager. Project management is a continuous education. Early success may be a detriment, as it may come too quickly and with not enough appreciation for the outcome. You will carry your head too high. I have seen proud individuals refusing to change because their pride becomes their identity. If challenged, it becomes personal rather than a healthy conflict of differing opinions. As I discussed in Chapter 9, two opposing opinions can be correct. Pride makes that statement untrue and must win to defend itself. Relinquishing such pride will help you become a better project manager and, in turn, a much better leader.

11.8. Social Intelligence

Social intelligence is defined as the ability to read people and understand silent cues. Those cues are body language, intent, inflection, and so on. People will say one thing but mean something different. If you do not have social intelligence, you will read these situations incorrectly, leading to preventable failures. Being swept up in your emotions leads to irrational decision making. For example, assume someone was working for you, and while his work was rational and realistic, he had trouble relating to people. Emotions would get the best of him. Projects are successful because of all the people involved with them, not just one person. Your social intelligence is important to project success. You have to handle situations with grace when anger seems to be the only emotion. Be a diplomat.

In relating back to the Von Manstein matrix (see Chapter 4), being technically savvy is perfect for operations. People are not as much of a concern as are the machines doing the work. As project managers, human interactions are the majority of your day. Whether it is calling a stakeholder, having a team meeting, or lunching with senior managers, you are dealing with people constantly. Social intelligence becomes intensified because of your daily interactions. A classic response to "How is your day?" is "Living the dream." If you have zero social intelligence and cannot recognize sarcasm, your belief is that person is living the dream. In actuality, that person could not be less happy. It is an obvious marker of something that needs improvement. If you are too caught up in the day-to-day work, you will miss these signs. Of course, some people can never be happy. On the other hand, some are truly unhappy, show signs of unhappiness, and life continues. No one sees they are unhappy. No one sees their performance has gone down. The team and project suffer. People will suggest the unhappy employee should step up and say something. What happens when that does not work? As a project manager, you have to observe and be proactive. See these situations coming and try to reach out as much as you can. Observation, as discussed in Chapter 7, is crucial to project success.

Robert Greene, in *Mastery,* says there are two components to social intelligence—specific and general knowledge of human nature (Greene, 2012). Specific knowledge is the ability to read people and understand their individuality, having a "Spidey Sense." This sense lends itself to a more fluid situation such as project teams changing, someone getting fired, bringing on new hires, and so on. General knowledge is accumulating an understanding of the overall patterns of human behavior that transcend us as individuals. This knowledge lends itself to more static situations. Examples are biases people bring with them to the decision-making process. The hot hand fallacy, sunk cost fallacy, framing bias, optimism bias, and the illusion of control are all biases people unconsciously exhibit while making decisions. Knowing these frameworks (Chapter 9) helps a project manager to realize the situation and make a change based on what another party is doing. It is similar to counter-punching in boxing. You are seeing how they react while you remain steps ahead of them. For every move or action they have, you have two or three defenses to counteract.

Often, work becomes busy, and tasks increase. Project managers get stuck with their heads in the sand. They do not see the forest for the trees. Social intelligence needs to be brought to the forefront when times are difficult. There will be obvious signs of distress. You will notice who handles these situations better than others. In the day-to-day activities, specific knowledge of human nature will provide the best results. You will know your team perhaps as well as your family. You will recognize situations for what they are not. People will exhibit biases; you will recognize them, and make the right decisions because you are socially aware of what is in front of you.

Chapter 12 discusses giving back. This charity does not have to be in a traditional sense of volunteering or charity work. While those options are acceptable, there are many different ways to give back as a project manager. The first avenue to give back is to yourself. Sounds selfish, but your happiness impacts those around you. I discuss the happiness equation and how to increase your happiness. Andrew Carnegie's *Gospel of Wealth* (1998) is discussed and how we can impact generations to come. Other topics discussed are giving recognition and how easy it is to do so, simply being human, and passing down the tools for others to succeed. Giving back does not have to be monetary. Chapter 12 discusses many forms of giving back to your organization and team with very little effort yet with a major impact.

Chapter 12

Giving Back

In *Bird by Bird*, Anne Lamott states, "If you give freely, there'll always be more" (Lamott, 1995). There is no cost for being nice and cordial. Back in eighth grade, my class went on a retreat. It was a team building exercise, getting to know your classmates graduation celebration. Part of that retreat was an activity based on being trapped on a snowy mountain. Each of us put on a blindfold (because the snow was a blizzard), stepped into a large maze layout, and had to find the exit off the mountain. Once you found the exit, you had two options. You could go back and help your classmates find the exit, or you could simply exit the maze. My strategy was to place one foot along the edge of the maze and shuffle my feet along and feel for the opening. At some point during this exercise, a counselor taps me on my shoulder, and I immediately exit the maze and take my blindfold off. After 10 or 15 minutes go by, the counselors end the game, everyone takes their blindfolds off, and they realize there was no "exit." The only way to get off the snowy mountain was to ask for help. Apparently, in my search for the exit, I must have asked for help. The experience still sticks with me because I never went to help any of my classmates. I exited selfishly but regretted doing so. The lessons learned that day were that asking for help is encouraged, and if someone does help you, return the favor.

When you decide to give back, realize that your time is more valuable than the check you write. Think about the times you have spent with family for vacations or holidays. Those memories could not be replicated with money. You can always spend money for a vacation, but it is the people who are with you that make the trip worthwhile. Writing a check and mailing it takes all of one minute to do. Spending an hour with someone or a group of people truly is a selfless act.

If you have had to take care of an elderly person, you know it is a meticulous process where help is necessary for most activities. Use this same approach when dealing with apprentice or assistant project managers. Giving back involves an exchange of knowledge as well. Go into it knowing the challenges you will face. Frustration will build, but do not let that discourage you from developing their skills. Treat them like a fine wine or a marinade: The longer it sits and does its thing, the better it will end. In working with the individual, you will know when it is time to let the bird fly. Kick the bird out of the nest and test skills. Inventor Dean Kamen says, "In a world of material goods and material exchange, trade is a zero-sum game. But if you have an idea and I have an idea, and we exchange them, then we both have two ideas." How often do we keep secret our way of doing things? Do we share industry "secrets" to enhance our team members? Build on them to make them even greater. I have been around too many senior managers who always have plans, but plans only they know. What good does it do to come up with a plan that only you know? Share that plan. Grow that plan. Add new ideas. Your plan may not have been grand enough once people start to talk about the possibilities. Sharing information is one of the easiest ways to give back. Who does not like to share what they know? Ask people about a passion of theirs and listen to them talk for hours. They will know every aspect of every piece. Lending an ear is an underrated form of giving back.

Happiness, not only others' but your own. Focus on it. Cultivate it. Grow it. Aristotle said that the goal of all human activity should be happiness. Perform activities that lead to your happiness. Many think that if they had a million dollars, they would be happy. I once asked a senior manager of mine, "If you had $10 million right now, what would you do with it?" His response was to buy a bigger house, build a pond in his backyard, and buy a luxury vehicle. I suggested he could have all of those things right now without the $10 million. The money is not the issue. It is the action that becomes the issue. Do what makes you happy, no questions asked.

Figure 12.1 represents a very simple decision that is too often forgotten. Pick projects or activities that will make you happy. Choose endeavors that will make you grow as a person. Having a positive attitude about the organization where you work and the projects you lead will encourage your team members to do the same. It will naturally give back to them. You will be more apt to teach and learn so as to allow others to grow and give back. Have you ever asked people about their hobby and seen their faces light up while they go on and on about the activity? They want to tell you everything. New technology is coming out. The latest program or set is about to hit the market. That is how you should be about your work: constantly keeping up with the latest and greatest.

Activities	 😀	 😢
Working out	✓	
Sports		✗
Leading people	✓	
Construction		✗
Taking dog for a walk	✓	

Figure 12.1 List of activities.

I found myself not acting in this manner. At that point, I knew I had to get leave. Walking toward the office door each morning was like walking the "green mile." I did not want to learn, and I did not want to teach anyone anything. As horrible as it is to admit, seeing them struggle made me happy. If they had a problem, I would wait for them to ask me instead of freely giving the needed information. Eventually, they would come to me for help, and I would assist them, but I never did it proactively. It was the most negative situation I have ever been a part of, and it made me realize how much happiness *needs* to play a role in my future.

12.1. Stoicism

What injures the hive injures the bee.

– Marcus Aurelius, *Meditations*

The bigger picture is the message instead of thinking about yourself and your actions. Think about how what you do will impact others around you. Your attitude toward the day and the project impact your team's attitude. You are the leader they look to for direction. If your confidence is fading, how can you expect them to be confident? There is a domino effect. Your single action ripples outward, affecting those around you. Even things you do not do have an impact. Not making that phone call can be worse than saying the wrong thing. Simply asking "How are you?" can have a major impact on a person's day. I remember walking into Accounts Receivable and having a woman ask "How are you doing?" with sincerity in her voice. It struck me. She cared how I was doing, and it gave me the chills. I am not used to that in the construction sphere. Most people are concerned about their projects and what moves need to be made next. It wasn't that they did not ask how I was doing, but it was the automated response. There was no purpose in their voice. It was something they felt they needed to say rather than wanted to say. Active listening needs to be stressed in giving back. Giving back is not just physical gifts, but also your time and attention. It is not just asking questions because you feel obligated, but wanting to know the answers and then remembering the answers. Then, bring up those answers the next time you talk to that individual. It goes a long way. It will start to ripple throughout your team. If people care about the people they work with, they will go above and beyond to help each other. That is the environment you want your hive residing in. And the second you take action to damage that environment, all the "bees" become negatively impacted.

12.2. Taoism

> *The Sage accumulates nothing.*
> *Having used what he had for others,*
> *He has even more.*
> *Having given what he had to others,*
> *What he has is even greater.*

– Lao Tzu, *Te-Tao Ching*

When my grandmother passed away, I donated many of her clothes and belongings to local shelters. It was the best feeling to walk into the room and see the people who would be affected by my donation. I considered it a small gesture, but these people were so thankful for the clothes, beds, dressers, tables, and so on. Keeping the furniture may have saved some money, but knowing I was contributing to a better place for these individuals was worth far more than the monetary gain I would have realized. Get your team involved outside of work activities. Make sure they give as much outside of work as they do at work. I have seen so many people give their all at work as they work late hours and stay away from home for weeks as they chase the money. Meanwhile, at home, things could not be worse. A company I worked for had signs posted in Human Resources about making sure you spend time with your family. Underneath that sign, it was about handling divorce. Obviously, it is an issue if there are signs posted around the office. Imagine the impact you could have if you gave the same effort you do at work as you do at home. I realize there are only 24 hours in a day, and work and sleep consume the majority of them, but use the few hours you do have at home to ensure things are harmonious. It may be a utopian dream of mine, but why not make an effort to get there?

Even if it is a simple phone call to an old friend to catch up, it is a plus. I have a friend who lives on the West Coast, and when I see his name pop up on my phone, the smile on my face goes from ear to ear. I always know it is going to be some crazy story from life on the road or catching up on Wisconsin sports. The impact of those ten-minute conversations far outweighs the ten-hour days at work. I could build the greatest spreadsheet in the world that tracks every aspect of aggregate production, but it does not compare to the draft breakdown for the upcoming football season. Giving back is more about using what you do at work to impact the people around at home positively. You can use the money you earned to throw the best holiday party your neighborhood has ever seen. Or, as another example, my mother always had time to go over homework after dinner. No matter the day she had, after dinner we would go over homework. At a certain point, she really would not help but instead provided the support of being there just in case. It is always more fun to do mundane tasks with someone else. Studying, reading, or preparing is better done in a group setting. If you sit down and try to read a specification book by yourself, your eyes will get heavy a few minutes later. But if you can drag someone else along, it always makes it better.

My dream is one day to start a scholarship fund at my alma mater. I always thought it was cool when people came back to their humble beginnings, whether it was college or where they grew up in the neighborhood. Success can be taken away as quickly as it can be earned. Giving back is a lasting way to build on your success. Doing what you do for others will give you that extra motivation when the chips are down, and no one believes in the project anymore. My plan is one day to give back and help the students who share the same passion as I do for the science of management. I never received a scholarship throughout college, but I can imagine the joy when the letter comes stating you have earned a scholarship. Part of that scholarship would entail being active in your community. While giving back to the local PMI chapter is my current contribution, it is just the beginning. When I had to move and stop volunteering at one chapter, they gave me more of a goodbye than a company where I had worked for years did. For me, that said something, considering I had volunteered once a month for little over a year. This is the impact you can have on others.

12.3. Sales

Make use of your "power base," as Grant Cardone labels it (Cardone, 2012). This power base is your network, the people you have dealt with over the years to get where you are. But instead of using them to get a business advantage, use them to help others. Auction items are a great way to use your network. I have participated in many silent auctions to raise money for kids in need or scholarships for college students. Seeing how generous some people can be is inspiring to me. My family would purchase a sponsored table for one event. It involved first getting access to get autographs from professional baseball players before the rest of the crowd. Those were experiences I will never forget. Not only giving the donation to the charity, but also seeing those players up close and having a few minutes with each of them to talk about baseball. It was a great way to use our resources for the greater good. After that, there would be silent auction items surrounding the hall that people could bid on. Seeing how generous the ballplayers were with their time made me realize how selfish I was with mine.

One of the players started the charity and was using his network to help hundreds of children each year. Using his money as well as donations throughout the year allowed these kids to have an escape from the daily turmoil they were facing. It takes a player one minute to sign a baseball, but the money generated from putting that pen to a ball can give people opportunities they never thought possible. That is powerful. Seeing the smiles on the children's faces while the video plays still gives me chills.

Even with writing this book, people have given me an opportunity I never thought possible. My small network of professionals made this possible, and I could not thank them more. I hope one day to pass a similar opportunity on to an aspiring professional. Not only to have the ability to give back to the profession, but also to bestow this opportunity on to someone else to advance the profession even more. It is something I once did not think would be possible. Not letting them down gives me the drive to keep typing and keep giving it my all.

"Safety and comfort are mortal danger to the soul," wrote Sam Sheridan in *A Fighter's Heart* (Sheridan, 2007). Sheridan goes on to write about how no good paintings or art came easily. They were all battles. Some things are worth fighting for and encourage you to put all your eggs in that one basket. This chapter is not just about giving back to those around you, but also leaving a legacy within your organization. Be someone who makes a difference and creates positive change. Safety and comfort within an organization lead to complacency and falling behind the times. Every day is a good day to make an impact on somebody or a group of people. I have seen senior managers come into the office, close their door, and not open it until lunch, and then repeat that same effort until it is time to go home. Their impact is being reduced by their selfishness. In my experience, open-door policies work best. They encourage people to ask questions, stop in to say hello, and have more fun while at work. Giving back is not just about money or time, it is about impacting someone's day-to-day life. Employees spend more conscience hours at work than at home. You will have a large impact on their lives. Make it positive and growing.

I was sitting in a PMI monthly meeting when the speaker (Ralph Nussbaumer, a project manager at AT&T) said, "One kind word when you lose is worth two when you win." This quote resonated with me. Construction beats people down physically and even mentally. If a mistake is made, yelling ensues. But those "losses" are a great time to teach and give words of encouragement. That does not mean raising your voice is never necessary. There is a time and place for everything. Make sure to counteract your way of handling the situation with a teachable moment and reflection. When the project is rolling, and the schedule looks good, it is easy to give pats on the back and words of praise. When the chips are down, and stakeholders are upset, pushing that negativity off onto our team can be easy, and it is a release of sorts. Giving back to your team in these moments of hardship will mean more then than when times are good. Showing your ability to care and showing compassion go a long way.

An example in construction is a death in the family. These occur unexpectedly. During the construction season (April–November), it is frowned on to take vacation time. We would have crew members come to us looking for a vacation day during the busy season because a family member had passed away. I could not believe the responses some project managers had. They showed no sympathy or compassion over the situation. They acted as if the crew member had done something wrong. How do you think that made the crew members feel? Do you think they wanted to come to their project managers with issues, even as serious as death? Meanwhile, the project managers were going on golf trips with clients and leaving early on Fridays. Again, a few kind words in a time of loss go a long way. Do what is difficult when it is difficult. Do not wait for someone else to break the ice and only then follow suit.

12.4. Happiness: It Is Simple Math

> *So there are many ways to be rich: You earn, inherit, borrow, beg, or steal enough money to meet all your desires; or, you cultivate a simple lifestyle of few desires; that way you always have enough money.*
>
> – Dan Millman, *Way of the Peaceful Warrior* (2000)

This quote put in simple mathematical terms is as follows:

$$\text{Happiness} = \frac{\text{satisfaction}}{\text{desires}}$$

There are two ways of impacting your happiness: Increase your satisfaction or decrease your desires. The ways in which to do so will depend on your subjective viewpoint. Only you know the activities and outcomes that create the most satisfaction for you. Maybe it is celebrating more small wins, or, to use a baseball analogy, instead of looking for the home run, realize that four singles get you the same outcome of scoring one run. Your perspective on your situation may need to change. Maybe things are not so bad, and if they are, move to a situation where your satisfaction will increase immediately. Do not get stuck in a rut.

The alternative to increasing your happiness is to lower your desires. This lowering of desires does not mean giving up on your goals or lowering your expectations. The equation relates to what you want versus what you need. Start to pick out the characteristics you need in a situation rather than all the wants. Your list of wants is probably greater than your needs. Create a mindset built around your needs. Physically write those needs down and begin to analyze. Does your current situation meet all of your needs? If not, who/what/where does possess all of the needs? What steps do you need to take to get there? Do the math. You now know satisfaction needs to increase or desires need to decrease to be happier. And who does not want to be happier?

12.5. Gospel of Wealth

Andrew Carnegie, an American steel industrialist who became one of the richest men in the world, wrote an article titled *The Gospel of Wealth* (1998). He described ways to use your wealth to benefit the most people in your community. Since he was the richest man of his time, he felt the need to help others while he was still living rather than letting his family inherit the money to do as they pleased. Carnegie mentions three modes of disposing of your surplus income:

- Left to the families of the decedents
- Bequeathed for public purposes
- Administered by its possessors during their lives

Carnegie thought the first option was more about family pride than about providing welfare for the family. Moderate allowances were acceptable to hand down to your family, but anything exceeding moderation was thought of as useless. The theme of moderation was present throughout the article. What that was would be determined by the public, because he felt the public has an idea of what is acceptably moderate.

The second choice, bequeathed for public purposes, was seen as unnecessary to Carnegie. If you want to help the public, why wait until you are deceased? Waiting is a fine option "provided a man is content to wait until he is dead before he becomes of much good in the world." He did not see a reason to wait to help out the community. If you have the means, put them to use in places you feel are important. If you give the money to someone else, and their views are different, they may use your money where you would not see fit.

The last choice, administered by its possessors during their lives, seemed most logical to Carnegie. In his support of giving during your life, he warns about how you give. He tells a story of a man giving a quarter to a beggar. The man knew nothing of the beggar's intentions but had every reason to doubt the beggar. Carnegie argues that the donation "will probably work more injury than all the money will do good which its thoughtless donor will ever be able to give in true charity." The donor did this to feel good about giving, but the donation will not result in the betterment of the beggar. It is similar to the discussion in Chapter 2 about having a passion with no purpose. This kind of giving is donating without a purpose. It goes into the ether. The effect is almost certainly negative. Carnegie states, "the best means of benefiting the community is to place within its reach the ladders upon which the aspiring can rise."

Carnegie's placing of ladders was in the form of libraries (Figure 12.2). These places would be for people aspiring to improve. You had to show initiative to learn and grow. This charity was not a monetary handout. It was a place for education. Other ways he extended ladders to the community were through parks, means of recreation, works of art, and public institutions. He wanted to help not only the body and mind, but also improve public aesthetics. Doing the most for the most people was his goal, such as improving the community as a whole instead of individually.

I mention the gospel of wealth because project managers have the ability to affect people positively. Our organizations, teams, and projects impact more people than the stakeholders we list in the register. There are generations to come who benefit from a wider freeway system, an application, or public institutions. Profits are important and so is impact. Making money while destroying

Figure 12.2 Carnegie Library.

the environment does not seem like a fair trade-off when contrasted to making money while freeing up a generation's time, money, or resources. Create this impact while you still can, and while you are in control.

12.6. Recognition

Few things offer greater return on less investment than praise—offering credit to someone in your organization who has stepped up and done the job.

– Bill Walsh, *The Score Takes Care of Itself*
(Walsh, Jamison, and Walsh, 2009)

Bill Walsh took his football team from 2 wins to 14 wins to the Super Bowl in three seasons. He was given a down-and-out team with little hope for the future and turned it into one of the greatest franchises in the National Football League. He was hard on his players, expected the world, and did not want excuses. Walsh gave them a standard of performance they had never seen before. While placing these great expectations on his players, he rewarded them for a job well done. He would single out players for performing poorly, but he would also single out players for doing their jobs. Think about how easy it is to tell someone he has done a good job. It is a simple sentence leaving your lips that takes no more than three seconds to utter. The payoff to those words is an encouraged employee wanting to stay a top performer. Deadlines are fast approaching. Budgets are thin. The pressure to perform is always there. A simple "Good job" can alleviate the stress of the job.

As a project manager, you are similar to the editor for the publisher. Hearing praise for one idea makes up for the paragraphs of red marks you received. There is always room for improvement. Best practices can become better. Your writing and presenting skills can be sharper. The way you handle conflict can improve. Hearing words of encouragement lets you and your team know you are on to something. You are doing something great. Those extra hours in the office sacrificing your free time become worth it when your manager tells you the presentation was excellent.

Employees may think getting paid more would make them happier at their job. Maybe. The issue could also be recognition or a feeling of importance. I talked about ego and trying to suppress the ego when making decisions. That does not mean it does not exist. For many of us, ego is very alive. Recognition feeds into positive aspects of the ego. Sending an email out to the team letting everyone know how well they are performing boosts morale and keeps the team on your side. Constantly pushing for more and more without giving recognition kills morale. No one wants to work for a dictator. Being a part of the team is as important as anything. Give credit where credit is due. Recognize your team for the hard work they have put in. Those words do not take much to say, yet the impact is tenfold.

12.7. Being Human

Do to others as you would have them do to you.

– Luke 6:31

The Golden Rule. We have heard it since we were young. As adults, though, we often forget this simple rule. Divisions of organizations become competitive against each other while working under the same roof, backstabbing each other because money and ego get in the way of doing right. Do not forget that we are in this together. The success of the organization depends on the success of its parts. Robbing from Peter to pay Paul. Biting the hand that feeds you. All of these phrases were developed because the actions they represent became commonplace.

While working for a large organization, there were numerous times daily when I would pass someone in the hallway without a word said. I would try to make eye contact, hoping to strike a quick hi–bye in the hallway. People would not even participate in that small activity. It was an act so innate yet so difficult. They were so busy with their heads buried in their phones that they did not have time to look up for a quick greeting as we passed in a hallway barely big enough for the two of us to squeeze through easily.

Going to lunch and picking up the check is a manager move and a strong one. There is nothing an employee appreciates more than a free lunch. The surprise on their faces. This relates to the recognition section above. The small investment of lunch pays off big down the road. People remember these small acts of kindness along the way. The idea is to do them from a place of goodness rather than the expectation of return. The return is natural. It comes with the territory. Being human is the golden rule. It is simple in words but complicated by life. An entrepreneurial project manager not only brings in profit for the organization and runs a well-oiled machine for a team but also returns the favor by recognizing the hard work put in by everyone involved and treating people with honesty, integrity, and respect.

12.8. Show Me the Tools

Stewart Brand, a writer and editor, recalled a conversation he had with Buckminster Fuller, the noted architect, systems theorist, author, designer, and inventor. Fuller spoke of human nature and how it has remained the same for a very long time. Instead of trying to change human nature, Fuller recommended going after the tools. Brand quoted Fuller as saying, "New tools make new practices. Better tools make better practices" (Diamandis and Kotler, 2012, p. 120). Project managers have to deal with the resources they are given. Your team will change when the project is complete. Instead of trying to change each person's abilities or capabilities, give them the tools to succeed. Put them in positions to win. If some team members are not technically savvy, give them a task in marketing, scheduling, estimating, and so on. Both parties will become frustrated if you place a nontechnical person in a technical role.

Coaching trees in sports are interesting to view. Many successful coaches have learned under other successful coaches. Bill Parcells, Hall of Fame NFL coach, is a great example of someone who passed tools down to his understudies. A few of the names included among those he coached are Bill Belichick, Tom Coughlin, and Sean Payton. All of these coaches have won Super Bowls a common link of having worked under Bill Parcells at some previous time. Other coaches with impressive coaching trees are Bill Walsh, Paul Brown, and Tom Landry. These coaches have won at the highest level in their industry, and they have passed down their tools for others to come up and have success. I have worked at places where leaving to better yourself is seen as a cop-out and your success elsewhere as a dig at your current organization. Instead, the approach should be like these coaches have. You are a successful leader; you teach others how to be successful, and then they branch out to realize that success. It is like a proud parent seeing his or her kids grow up.

Working in fear of losing your job to someone you are teaching is an insecurity that does not belong in the workplace. In construction, that was a constant fear. It was as if people were keeping secrets from you to inhibit your ability to perform, so they would stand out from everyone else. A selfish team environment will never lead to success. A project manager is similar to a coach of a team. You have to pass down your information so the team can perform at its best. Why would a coach withhold information about the opponent? That is what happens when a project manager keeps everything hush. You are only hurting your team. Their success becomes your success. Jerry Rice, Steve Young, and Joe Montana all credit Bill Walsh for their success. These are some of the greatest football players ever to play the game. They had natural ability, but it took a great coach to maximize their potential. The same goes for project teams. All of your team members have talent. Otherwise, you would not have hired them. It is your job to realize that talent and pull it out of them. Give back by giving them the tools to succeed.

References

Adair, Red. 2016. BrainyQuote.com, Xplore Inc., https://www.brainyquote.com/quotes/quotes/r/redadair195665.html. Accessed December 21, 2016.

Aurelius, Marcus, and Gregory Hays. 2002. *Meditations.* New York: Modern Library.

Blanda, George. 2016. Pinterest, https://www.pinterest.com/pin/510595676483144428. Accessed November 2, 2016.

Brownlee, Dana (Producer). 2015. "4 Common Rookie Project Manager Mistakes and How to Avoid Them" [video webinar], http://www.projectmanagement.com/videos/303332/4-Common-Rookie-Project-Manager-Mistakes-and-How-to-Avoid-Them. Accessed September 3, 2016.

Cardone, Grant. 2011. *The 10x Rule: The Only Difference Between Success and Failure.* Hoboken, NJ: John Wiley & Sons.

Cardone, Grant. 2012. *Sell or Be Sold: How to Get Your Way in Business and in Life.* Austin, TX: Greenleaf Book Group Press.

Carnegie, Andrew. 1998. *The Gospel of Wealth.* Bedford, MA: Applewood Books.

Cooper, Lawrence (Producer). 2016. "Organizational Agility—We Asked Questions. We Got Answers" [video webinar], http://www.projectmanagement.com/videos/340994/Organizational-Agility---We-Asked-Questions--We-Got-Answers. Accessed August 16, 2016.

Covey, Stephen M. R., and Rebecca R. Merrill. 2006. *The Speed of Trust: The One Thing That Changes Everything.* New York: Free Press.

De Flander, Jeroen (Producer). 2016. "Uncover 3 Hidden Change Levers to Successfully Execute Your Strategy" [video webinar], http://www.projectmanagement.com/videos/343787/Uncover-3-Hidden-Change-Levers-to-Successfully-Execute-Your-Strategy. Accessed September 7, 2016.

DeBono, Edward. 2000. *Six Thinking Hats.* London: Penguin Books.

Denning, Stephen. 2010. *The Leader's Guide to Radical Management: Reinventing the Workplace for the 21st Century.* San Francisco, CA: Jossey-Bass.

Diamondis, Peter H., and Steven Kotler. 2012. *Abundance: The Future Is Better Than You Think.* New York: Free Press.

Dweck, Carol S. 2006. *Mindset: The New Psychology of Success.* New York: Random House.

Dyer, John R. G., Darren P. Croft, Lesley J. Morrell, and Jens Krause. 2009. Shoal composition determines foraging success in the guppy. *Behavioral Ecology,* 20, 165–171.

Elatta, Sally (Producer). 2016. "Adaptive Leaders—Assessing and Growing Your Agile Leaders" [video webinar]. http://www.projectmanagement.com/videos/342125/Adaptive-Leaders---Assessing-and-Growing-Your-Agile-Leaders. Accessed August 24, 2016.

Emerson, Ralph Waldo. 2016. BrainyQuote.com, Xplore Inc., https://www.brainyquote.com/quotes/quotes/r/ralphwaldo101322.html. Accessed November 30, 2016.

Fajans, Jeff. 2014. "How to Become a Creative Badass: A 9 Step Guide to Mastering the Creative Process." Create.Learn.Live, http://www.createlearnlive.com/blog/2014/10/6/how-to-become-a-creative-badass. Accessed October 6, 2016.

Fibonacci spiral. 2016. Pixabay. https://pixabay.com/en/fibonacci-spiral-science-golden-1601158. Accessed December 21 2016.

Fosbury, Dick. 2016. "Fearless Fosbury Flops to Glory." http://www.nytimes.com/packages/html/sports/year_in_sports/10.20.html. Accessed November 3, 2016.

Gates, Bill. 2016. BrainyQuote.com, Xplore Inc., https://www.brainyquote.com/quotes/quotes/b/billgates104353.html. Accessed November 30, 2016.

Gogolak, Pete. 2016. http://alchetron.com/Pete-Gogolak-572432-W. Accessed November 3, 2016.

Gordis, Leon. 2009. *Epidemiology*, 5th ed. Philadelphia: Elsevier/Saunders.

Greene, Robert. 2012. *Mastery*. New York: Viking.

Harpham, Bruce (Producer). 2016. "How to Play Office Politics Without Losing Sleep" [video webinar], https://www.projectmanagement.com/videos/344450/How-To-Play-Office-Politics-Without-Losing-Sleep. Accessed September 14, 2016.

Hiatt, Sean. 2016. "A Brief History of Wallace, Idaho," Spokane Historical, http://spokanehistorical.org/items/show/485. Accessed December 21, 2016.

Holiday, Ryan. 2016. *Ego Is the Enemy*. New York: Portfolio, Penguin.

Hurwitz, Marc, and Samantha Hurwitz (Producers). 2015. "Leadership Is Half the Story: A Fresh Look at Followership, Leadership, and Collaboration" [video webinar], http://www.projectmanagement.com/videos/296695/Leadership-is-Half-the-Story--A-Fresh-Look-at-Followership--Leadership--and-Collaboration. Accessed June 16, 2016.

Juran, Joseph M., Frank M. Gryna, and Richard S. Bingham. 1974. *Quality Control Handbook*. New York: McGraw-Hill.

Klein, Gary A. 2004. *The Power of Intuition: How to Use Your Gut Feelings to Make Better Decisions at Work*. New York: Currency/Doubleday.

Knight, Bobby. 2013. *The Power of Negative Thinking: An Unconventional Approach to Achieving Positive Results*. Boston: New Harvest, Houghton Mifflin Harcourt.

Koch, Richard. 1998. *The 80/20 Principle: The Secret of Achieving More with Less*. New York: Currency.

Kreider, Tim. 2012. *We Learn Nothing: Essays and Cartoons*. New York: Free Press.

Lamott, Anne. 1995. *Bird by Bird: Some Instructions on Writing and Life*. New York: Anchor Books.

Lincoln, Abraham. 2016. BrainyQuote.com, Xplore Inc., https://www.brainyquote.com/quotes/quotes/a/abrahamlin109275.html. Accessed December 21, 2016.

Mandino, Og. 1968. *The Greatest Salesman in the World*. New York: F. Fell.

Maxwell, John C. 1998. *The 21 Irrefutable Laws of Leadership: Follow Them and People Will Follow You*. Nashville, TN: Thomas Nelson.

Michalko, Michael. 2006. *Thinkertoys: A Handbook of Creative-Thinking Techniques*, 2nd ed. Berkeley, CA: Ten Speed Press.

Millman, Dan. 2000. *Way of the Peaceful Warrior: A Book That Changes Lives*. Tiburon, CA: H. J. Kramer.

Missouri Department of Transportation. 2005. MSE Wall 04, http://www.modot.org/business/standard_drawings/documents/MSE_Wall_04.pdf. Accessed July 7, 2016.

Nielson, Jakob. 2006. "F-Shaped Pattern for Reading Web Content." Nielson Norman Group, https://www.nngroup.com/articles/f-shaped-pattern-reading-web-content. Accessed October 13, 2016.

Norton, Michael. 2016. "There's a Right Way to Do a 'to-Do' List." *Highlands Ranch Herald*, September 2, 2016. Accessed September 4, 2016.

Optical illusions. 2016. BrainDen, http://brainden.com/optical-illusions.htm. Accessed October 12, 2016.

Paul, Jim, and Brendan Moynihan. 2013. *What I Learned Losing a Million Dollars*. New York: Columbia Business School.

Provoost, Sandrine (Producer). 2016. "Preventing the 10 Common Mistakes in Leading Your Change and Transformation Projects" [video webinar], http://www.projectmanagement. com/videos/332245/Preventing-the-10-common-mistakes-in-leading-your-Change-and-Transformation-projects. Accessed June 2, 2016.

Quiring, Collin (Producer). 2014. "Organizational Project Management Maturity Is Project Management Quality" [video webinar], http://www.projectmanagement.com/videos/286193/ Organizational-Project-Management-Maturity-IS-Project. Accessed April 30, 2016.

Scissor approach. 2016. https://commons.wikimedia.org/w/index.php?curid=1815925. Accessed November 2, 2016.

Sheridan, Sam. 2007. *A Fighter's Heart: One Man's Journey Through the World of Fighting*. New York: Atlantic Monthly Press.

Sims, Peter. 2011. *Little Bets: How Breakthrough Ideas Emerge from Small Discoveries*. New York: Free Press.

Street, Farnam. 2014. "Why Clever and Lazy People Make the Best Leaders," *Business Insider*, March 14, 2014, http://www.businessinsider.com/why-clever-and-lazy-people-become-leaders-2014-3.

Syrus, Publilius. 2016. AZQuotes.com, Wind and Fly Ltd., http://www.azquotes.com/quote/ 951076. Accessed December 21, 2016.

Taleb, Nassim Nicholas. 2007. *The Black Swan: The Impact of the Highly Improbable*. New York: Random House.

The New York Stock Exchange|NYSE. 2016. "The New York Stock Exchange," https://www. nyse.com. Accessed September 15, 2016.

Tzu, Lao, and Robert G. Henricks. 1989. *Te-Tao Ching: A New Translation Based on the Recently Discovered Ma-Wang-Tui Texts*. New York: Ballantine.

Waitzkin, Josh. 2007. *The Art of Learning: A Journey in the Pursuit of Excellence*. New York: Free Press.

Walsh, Bill, Steve Jamison, and Craig Walsh. 2009. *The Score Takes Care of Itself: My Philosophy of Leadership*. New York: Portfolio.

Wheeler, John Archibald. 2016. BrainyQuote.com, Xplore Inc., https://www.brainyquote.com/ quotes/quotes/j/johnarchib201710.html. Accessed December 21, 2016.

Wright, Eric (Producer). 2016. "Making Better Project Decisions More Often" [video webinar], https://www.projectmanagement.com/videos/330460/Making-Better-Project-Decisions-More-Often. Accessed May 16, 2016.

Index